of the child that was me. –Judy Collins

Making Decorative
Mirrors & Shelves

Making Decorative
Mirrors & Shelves

Holly Jorgensen

Sterling Publishing Co., Inc. New York
A Sterling/Chapelle Book

For Chapelle Ltd.

Owner
Jo Packham
Editor
Karmen Potts Quinney
Designers
Holly Fuller
Sharon Ganske
Kelly Henderson
Susan Laws
Pauline Locke
Jamie Pierce
Cindy Rooks
Photographer
Kevin Dilley for Hazen
Photography Studio
Photography Stylists
Susan Laws
Jo Packham

Staff
Marie Barber
Malissa Boatwright
Kass Burchett
Rebecca Christensen
Marilyn Goff
Michael Hannah
Amber Hansen
Shirley Heslop
Holly Hollingsworth
Susan Jorgensen
Ginger Mikkelsen
Barbara Milburn
Linda Orton
Rhonda Rainey
Leslie Ridenour
Cindy Stoeckl

Several projects shown in this publication were created with the outstanding and innovative products developed by the following companies: Plaid, Delta, Aleene's, and Loctite.

We would like to offer our sincere appreciation to these companies for the valuable support given in this ever changing industry of new ideas, concepts, designs, and products.

Due to the limited amount of space available, we must print our patterns at a reduced size in order to give our patrons the maximum number of patterns possible in our publications. We believe the quality and quantity of our patterns will compensate for any inconvenience this may cause.

Library of Congress Cataloging-in-Publication Data

Jorgensen, Holly.
Making decorative mirrors & shelves / Holly Jorgensen.
p. cm.
Includes index.
"A Sterling/Chapelle Book"
ISBN 0-8069-9339-1
1. Handicraft. 2. Decoration and ornament. 3. Mirrors
4. Picture frames and framing. 5. Shelving (Furniture) I. Title.
TT157.J67 1997
745.593—dc21 97-33684
CIP

10 9 8 7 6 5 4 3 2 1

Published by Sterling Publishing Company, Inc.,
387 Park Avenue South, New York, NY 10016

Distributed in Canada by Sterling Publishing
% Canadian Manda Group, One Atlantic Avenue, Suite 105,
Toronto, Ontario, Canada M6K 3E7
Distributed in Great Britain and Europe by Cassell PLC, Wellington
House, 125 Strand, London WC2R 0BB, England
Distributed in Australia by Capricorn Link (Australia) Pty Ltd., P.O.
Box 6651, Baulkham Hills, Business Centre, NSW 2153, Australia
Printed in Hong Kong

Sterling ISBN 0-8069-9339-1

If you have any questions or comments or would like information on specialty products featured in this book, please contact: Chapelle Ltd., Inc.,
P.O. Box 9252 Ogden, UT 84409
(801) 621-2777 • FAX (801) 621-2788

Every effort has been made to ensure that all the information in this book is accurate. However, due to differing conditions, tools, and individual skills, the publisher cannot be responsible for any injuries, losses, and other damages which may result from the use of the information in this book.

Table of Contents

General Instructions 6-13

Chapter One: Frames 14-45
- Thimbles & Spools 14
- Play Ball 16
- Metallic Chain & Ball 18
- Oak Vanity 19
- Pearls & Lace 22
- Around the World 23
- Metal Moose 25
- Window Screen 26
- Square Tiles 28
- Bird's Nest 30
- Garden Wall 31
- Lighthouse 34
- Seashore 37
- Home Tweet Home 38
- Roses & Vines 42
- Teacher's Busy 43

Chapter Two: Techniques on Mirrors 46-65
- No Place Like Home 46
- Moonrise 48
- Family Heirloom 51
- Gilded Window 53
- Beveled Glass 55
- Scroll Saw Silhouette 57
- Floral Mirror 60
- Sundial 63

Chapter Three: Unique Mirror Ideas 66-89
- Garden Gate 66
- Fly Fishing 68
- Checkerboard Mirror 70
- Spiked Mirror 72
- Bunny Mirror 74
- Chair Vanity 77
- Blooming Mirrors 79
- Antique Balusters 81
- Window Box Garden 83
- Barndoor 85
- Pocket Mirrors 87

Gallery of Mirrors 90-96

Chapter Four: Shelves 97-120
- Checkered Heart 97
- Peach & Pear Shelves 99
- Blanket Shelf 102
- Baluster Shelves 105
- Life, Love, & Light Shelves 106
- Knickknack Shelf 107
- Moulding Shelf 110
- Plaid Heart Shelf 112
- Octagon Shelf 114
- Green Picket Shelf 116
- Copper Shelves 118

Gallery of Shelves 121-126

Metric Eqivalency Chart 127

Index 128

General Instructions

Before beginning thoroughly read the general instructions. Also, before using any tools or materials listed in this book, read and follow manufacturer's instructions. The colors used in this book are simply suggestions. Substitute any color(s) desired.

Gluing

Craft Glue

Craft glue is best for many projects. It is thick and all-purpose. It dries quickly, clear, and flexible.

Découpage Glue

Découpage glue is a glue, sealer, and finish all-in-one. Découpage glue comes in a gloss or matte finish. It is quick-drying, dries clear, and can be sanded to a smooth finish.

When découpaging, make certain surfaces are smooth and clean. Apply thin, even layer of découpage to small section of surface, using an old paintbrush or if preferred, a sponge brush. Press paper or fabric firmly onto découpage and rub gently, but firmly, to remove air bubbles and wrinkles. Continue découpaging until entire surface is covered. Dry overnight. Apply second coat of découpage over entire surface. Allow to dry until surface is clear.

Hot Glue Gun and Glue Sticks

Hot glue dries quickly and is available in clear, cloudy, or glitter sticks. Use the "cloudy" glue sticks when working with fabric. Clear glue sticks do not penetrate fabric well.

Use tweezers or needle-nose pliers to hold small objects in place. For larger objects, use craft stick or pencil to apply pressure until glue is hard. Strings of glue will be present, but are easily removed when hard.

Industrial-strength Glue

Industrial-strength glue is very strong and seals well. Use when nothing else will work. It is extremely toxic; use only in a well-ventilated area.

Tacky Glue

Tacky glue is used for applying fabric to crafts. Its sticky texture holds lightweight objects in place until dry.

Wood Glue

Wood glue is best for bonding wood. It is strong as well as heat and water resistant. Wood glue should only be used on oil-free surfaces. Pieces should fit snugly and a clamp should be used while drying for best results.

Hanging Mirrors

There are a variety of ways to mount mirrors. Items used to mount mirrors should be placed on frames before painting or embellishing. Mirrors in this book use two methods. The first way requires: eye screws, nails, and wire. The second way requires: hammer, nails, and one or two saw-tooth hangers.

Eye Screw Hangers

1. Place eye screws ⅓ of the way down on the frame. If an eye screw is placed any lower on the frame, the mirror will droop away from wall.

2. Cut needed amount of wire. Tighten each end of wire around eye screw.

Saw-tooth Hangers

1. Place saw-tooth hanger(s) on top backside edge of frame. Two saw-tooth hangers may be required for heavy mirrors. Attach following manufacturer's instructions.

*Shelves can be hung in the same manner as mirrors.

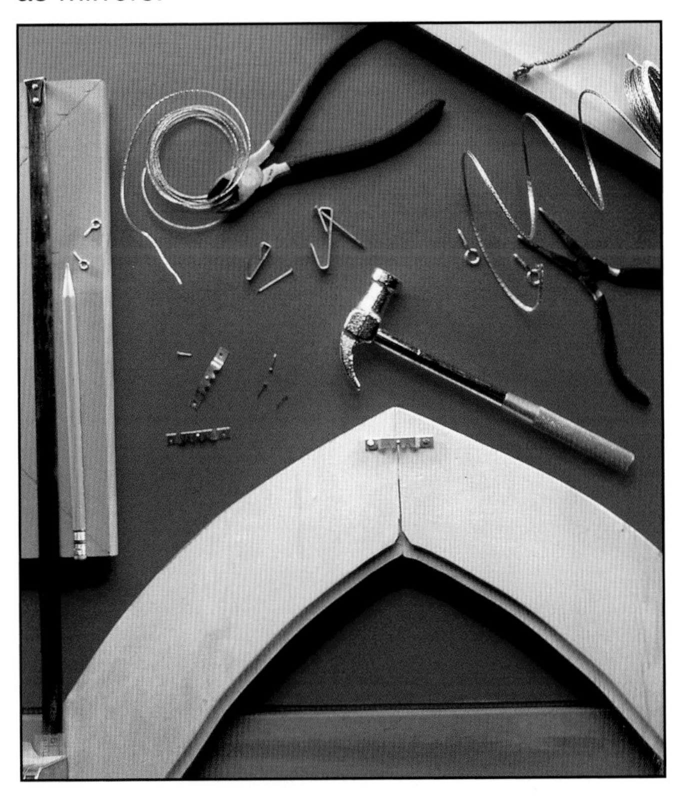

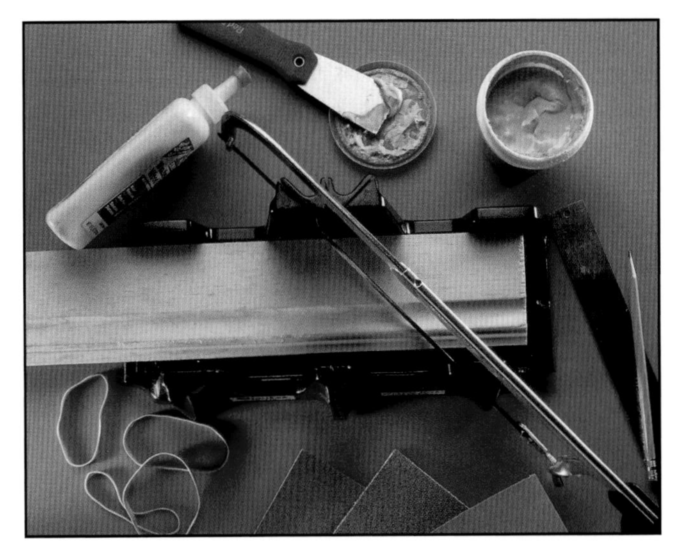

1. Cut four pieces of wood or moulding to desired length. Miter each end to 45° angle.

2. Glue joints together using wood glue. Secure frame with rubber bands. Let dry.

3. Fill any gaps with wood putty. Sand. Seal with acrylic gesso.

Making a Basic Frame

To make a basic frame the following materials are needed: acrylic gesso, large rubber bands, miter box, moulding or wood in desired size, sandpaper, saw, wood glue, and wood putty.

Paintbrushes

Paintbrushes are the most common tool used for painting. Good quality synthetic brushes work best when using acrylic paints. Paintbrushes come in a variety of different sizes. The size of the brush will depend on the size of pattern being painted.

Brushes are numbered. This refers to the shape of the metal part of the brush that holds the bristles onto the shaft. The higher the number, the larger the brush.

Be certain to clean brushes thoroughly with soap and water until the water runs clean.

Flat

Flat brushes have longer hairs and a chisel edge for stroke work. They are good for filling in, blending, and wide sweeping strokes.

A. Decorative.

B. Flat Basecoater. The large size bristles are for precise coverage of large areas.

C. Flat Scrubber. Helps paint and fabric mediums smoothly permeate fibers.

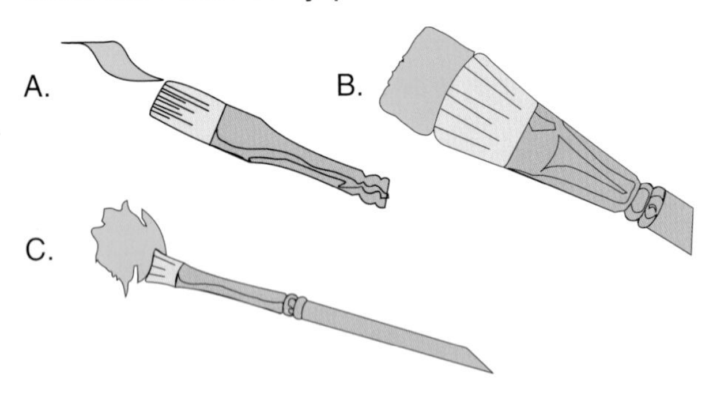

Liner

Liner brushes have a fine point and are good for delicate lines, detail work, lettering, and long continuous strokes.

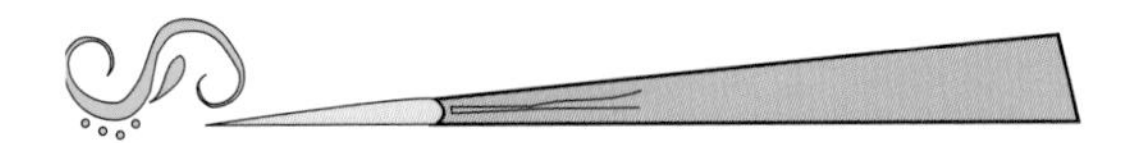

Spouncer

Spouncer brushes come in ¾" to 1¾" width sponges. They are good for pouncing, stippling, or swirling. They can be re-used when wet.

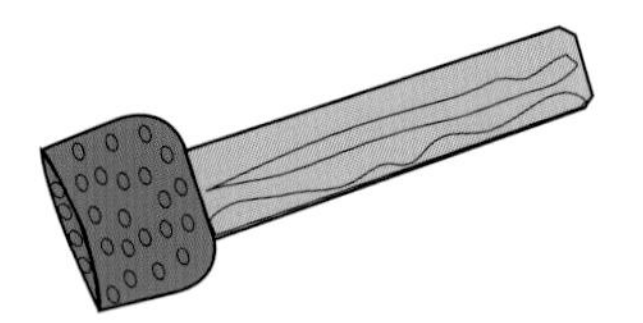

Paint Mediums

Antique Medium

Antique medium gives an old-fashioned, aged look to new paint.

1. Apply by rubbing over project surfaces with clean cloth or paintbrush.

2. Wipe off any excess. Allow to dry thoroughly. For additional information, read and follow manufacturer's instructions.

Crackle Medium

Crackle medium gives a weathered appearance to new paint.

1. Using acrylic paint, apply basecoat using flat brush. Allow basecoat to dry thoroughly.

2. Using old flat or round brush, apply one coat of crackle medium using long sweeping strokes; thin coat for small cracks, thick coat for large cracks. *Note: Topcoat will crack in direction crackle medium is painted on.* Allow crackle medium to dry thoroughly. Apply topcoat of contrasting color acrylic paint using flat or round brush or sponge. *Note: Topcoat must be applied for cracking to occur.* For additional information, read and follow manufacturer's instructions.

Painting Techniques

Basecoat

In most cases, projects in this book require entire surface to be painted with a basecoat.

A sealer, acrylic gesso, should be used before applying acrylic paint basecoat.

1. Apply acrylic paint to all surfaces.

2. Cover area with two to three smooth, even coats of paint. *Note: Apply several thin coats of paint, rather than one heavy coat. Lightly sand painted surfaces before applying additional coats.*

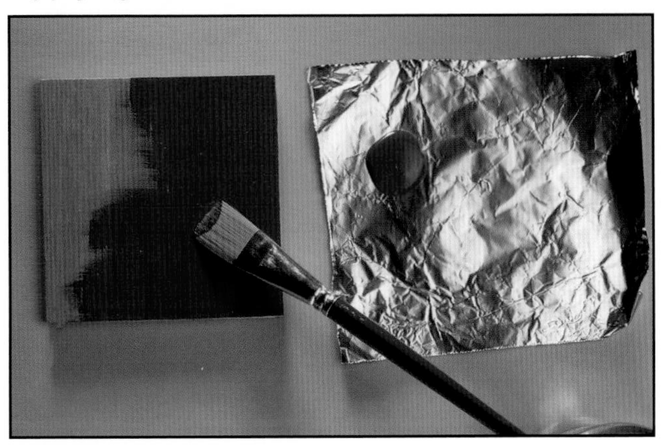

Basket Weave Stroke

1. Load paintbrush with colors.

2. Make square the width of brush with two horizontal strokes. Make next square with two vertical strokes. Repeat this process creating a row.

3. Start next row on opposite stroke of first row.

Checkerboard Stroke

1. Draw horizontal parallel lines spaced as desired over area to be painted.

2. Load paintbrush with first color. Stroke from first line to second line and lift paintbrush, creating one square. Space over equal width as previous stroke. Stroke and lift to create second square. Repeat to finish row.

3. On next row, space over width of paintbrush stroke. Stroke from second line to third line and lift paintbrush, creating one square below unpainted square. Repeat to fill area.

4. Load paintbrush with second color. Stroke and lift paintbrush filling in remaining squares.

Dry-brush

1. Using an old flat paintbrush, dip in a small amount of acrylic paint. Remove excess paint from brush by working in criss-cross motion on paper towel.

2. Using criss-cross motion, brush project with little to no pressure to create soft texture.

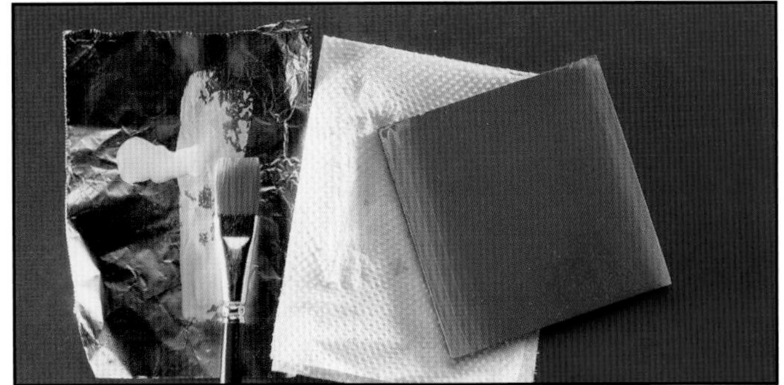

Float

1. Dampen largest flat paintbrush that will accommodate area to float. Wipe excess water on paper towel.

2. Load one corner of brush, up to ⅓ of width, of chisel edge of bristles with paint. Stroke brush back and forth on palette to work paint into bristles and soften color.

3. Apply brush to painting surface. Paint color should appear darkest at loaded corner and gradually fade to clear water on opposite corner. *Note: If paint spreads all the way across chisel edge of bristles, rinse brush and reload.*

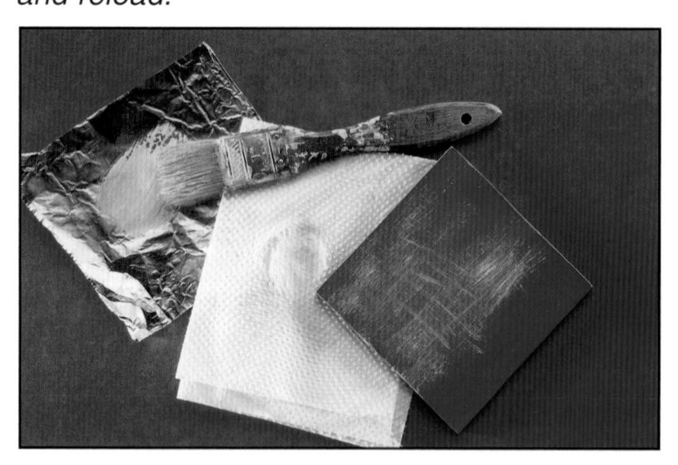

Marbleize

1. Basecoat project surface by loading flat or round brush with three to four different colors, blending slightly using one color as dominant color.

2. To make veins, dip paintbrush into one-part water and one-part paint. Pick a point on edge of project and lay brush down. Pull brush along surface with a twisting, turning motion, making a vein. The twisting motion varies the thickness, and the turning makes a crooked, natural looking line. Some parts of the veins should have more paint than others. Veins in marble are much like the branches of a tree – irregular, splitting, and often forming a Y-shape.

3. Repeat using a small, round liner paintbrush to make smaller veins.

Spatter

1. Using old toothbrush, dip bristles of toothbrush into paint that has been slightly diluted with water.

2. Hold toothbrush about 6 to 8 inches away with bristles pointed toward the project. Draw finger or thumb across bristles causing paint to spatter onto project.

Sponge Stencil

1. Place stencil on area to be painted.

2. Load sponge with very little paint. Sponge tip on paper towel to remove excess paint Apply lightly in area to be stenciled.

Stipple

1. Load brush or sponge with very little paint.

2. Bounce brush or sponge tip on paper towel then apply lightly to project. Vary dot sizes to create shadow or texture effect.

Swirl

1. Place small amount of each paint on aluminum foil.

2. Dip flat or round brush in each color. Swirl slightly on project with brush, building up paint to add texture. Do not swirl too much or colors will mix together, forming gray-brown color.

Wash

1. Mix one-part paint and three-parts water.

2. Apply paint wash to sealed wood using an old flat brush. Several coats of light wash will produce soft, but deep, transparent color. Allow wash to dry thoroughly between coats.

Sculpting Clay

Sculpting clay has excellent handling qualities. It is soft and pliable, and can be baked in a home oven. It takes tooling (drilling, sanding, and painting) well and can be rolled and cut.

Sculpting clay is available in a variety of colors or white clay may be used then painted.

Materials needed include acrylic gesso, glue of choice, knife, oven to bake clay, paintbrush, plastic wrap, and toothpicks or wire.

1. Knead clay until soft and smooth. Mold and sculpt clay into desired form.

2. Use paintbrush end, tooth-pick or wire, and knife to assist in making indentations, score lines, paint lines, and, if needed, to help support clay during sculpting.

3. Place clay form on wax paper and bake following manufacturer's instructions.

4. Paint clay pieces as desired.

Securing a Mirror

Flat Frame

To secure a mirror to a frame, gently place mirror inside or on top of frame. Use screen clips to secure mirror. Flat frames can also be secured with silicone glue following manufacturer's instructions.

1. Place mirror flat on back of frame. Run a thick bead of silicone glue along edge of frame. Smooth as needed. Let dry.

2. Mirror and back of frame should be covered with brown paper. This will protect back of mirror. Cut paper to fit back of frame. Run a bead of craft glue around edge of frame. Place paper on glue. Smooth edges of paper. Let dry.

3. When glue is dry, mist paper with water. Paper will tighten as it dries.

Routed Edge Frame

A mirror can be secured in a routed edge frame with glazer's points or small nails.

1. Place mirror face down in opening. Cover mirror with mat board or foam core that has been sized to fit opening. This fills gaps and protects back of mirror.

2. Using putty knife, gently push glazer's points into frame edges.

Soldering

A face mask should be worn while soldering to avoid breathing toxic fumes. If this is the first time using a soldering iron, practice on scraps. A little solder goes a long way.

1. Wrap copper foil tape around glass edges.

2. Using old paintbrush, apply flux over copper foil tape. Using old rag, wipe excess flux off glass.

3. Using soldering iron, solder around glass edges until desired look is achieved. Build up areas or make edges smooth or bumpy. If a mistake is made, touch iron to solder. The heat will melt solder again and allow its removal.The soldering spot can be covered with a variety of metallic paints that will blend with metal almost perfectly. Soldering may cause metal to tarnish. If this occurs, use fine steel wool to gently clean tarnished spot.

Stitches

Circular Ruffle

1. Fold ribbon in half, matching cut ends. Seam cut ends.

2. Gather-stitch along selvage edge. Pull gathers as tightly as possible and secure thread.

3. Completed Circular Ruffle.

1

2

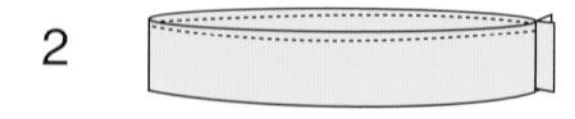

3

Running Stitch

1. Sew a line of straight stitches with an unstitched area between each stitch.

2. Come up at 1 down at 2.

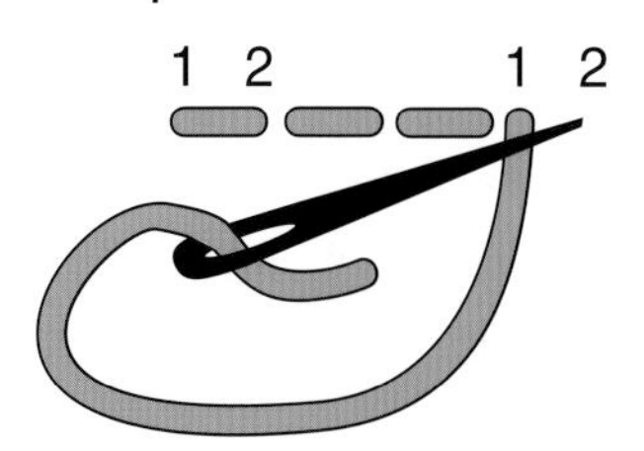

Tea Dyeing

1. Bring to boil, six to eight tea bags in two quarts of water. Turn off heat and steep for at least 20 minutes.

2. Place fabric in tea and soak for at least 30 minutes. The longer fabric is left in tea, the darker the color.

3. Rinse fabric slightly, ring out, and let dry.

Transferring Patterns

Transferring with Paper

1. If directions indicate to enlarge pattern, place pattern directly in photocopy machine. Enlarge required percentage.

2. Slip piece of transfer paper, graphite side down, between photocopy and material to be cut. Using pencil, trace outline(s) of shape(s) to be cut.

3. Cut out shapes.

Transferring from Grid

1. On 1" dressmaker's grid paper, mark dots on grid where pattern in book intersects its corresponding grid line.

2. When all dots have been marked, connect dots to finish pattern. *Note: instead of purchasing dressmaker's grid paper, make grid paper by drawing horizontal and vertical lines, spaced 1" apart on paper large enough to accommodate pattern at actual size.*

3. Transfer shape onto material to be cut.

4. Cut out shapes.

Tricks & Tips

1. Seal all wood with acrylic gesso. Sealing wood helps prevent wood grain from raising.

2. After the frame is built or purchased, decide where top and bottom of frame will be.

3. Attach hangers for mirrors and shelves before painting or embellishing.

4. When painting a mirror, always carry base color or design around to back of the inside lip as this will reflect in the mirror. Any piece that will be attached to the front of the mirror needs to be painted on all sides, as it reflects in the mirror.

5. Spray all painted surfaces with sealer.

6. Wash mirror before securing mirror into frame. Mix one teaspoon of rubbing alcohol and a pint of water in spray bottle for washing mirrors.

7. Arrange all embellishments in desired position before actually gluing.

8. Use masking tape to square off sections and to keep lines straight.

9. Clean gilded mirror frames with beer. Pour beer onto a soft rag, rub mirror gently, and wipe dry.

Chapter One

Frames

Thimbles & Spools

Thimbles & Spools

Materials

- Antiquing medium
- Bobbin: wooden, old
- Embellishments: assorted charms, buttons, sewing notions
- Fabric (1½ yds.)
- Fleece (1 yd.)
- Floss: assorted colors (7-8)
- Frame: octagon, 28" x 34" x ½"
- Heart: small, stuffed
- Lace: assorted doilies, tatting
- Measuring tape: cloth
- Mirror: cut to fit frame
- Ribbon: mesh, ⅝"-wide, (2½ yds.)
- Scissors: old
- Spools: wooden, assorted (15)
- Tea bags (4)
- Thimbles (2)

Tools & Supplies

- Glass bowl
- Glue: tacky; hot glue gun and glue sticks
- Paintbrush: stiff

Instructions

Refer to General Instructions on pages 6-13.

1. Cover front of frame with two layers of fleece. Cut to fit and hot-glue in place.

2. Tear fabric in 2" x 9" strips. Hot-glue end of one strip to back of frame, wrap once and secure other end to back overlapping slightly. Repeat wrapping process until entire frame is covered.

3. Use **Antique Medium** to antique embellishments, spools, thimbles, buttons, scissors, etc. Apply with stiff paintbrush. Let dry.

4. **Tea-dye** lace, doilies, tatting, and measuring tape. Let dry.

5. Wrap floss around spools. Secure ends with tacky glue.

6. Make looped bow with long tails from mesh ribbon.

7. Embellish frame as desired by hot-gluing items at random.

8. Complete by **Securing Mirror** into frame.

Faye Moskowitz

" Recently my mother has begun to come back to me in dreams. In the dream my mother caresses my arm with strong fingers that are twins of my own. She calls me 'faygeleh,' 'mammeleh,' and little daughter'–'tochterel.' I can hear her voice. She says, 'You have been a good mother. You did what you could, and so did I.' I forgive my mother for everything, as I pray my daughters will forgive me. When they look in their mirrors one day and my face appears, I hope they will love what they see."

Play Ball

Play Ball

Materials
- Acrylic gesso
- Acrylic paints: black, brown, green, purple, yellow
- Chain: small links, black (2')
- Eye screws: 11⁄16", gold (6)
- Mirror: cut to fit frame
- Wood balls: 2" (1); 1¾" (2)
- Wood frame: basic, 11" square
- Spray sealer: matte

Tools & Supplies
- Awl
- Glue: industrial-strength
- Masking tape
- Paintbrushes
- Pliers
- Sandpaper: fine grit
- Wire cutters

Instructions
Refer to General Instructions on pages 6-13.

1. Make a Basic Frame or use purchased frame.

2. Seal frame with acrylic gesso.

3. Square-off corners of wood frame with masking tape. Paint corners of frame with purple paint. Use yellow paint for top and bottom squares. Paint side squares green. See Diagram A.

Diagram A

4. Using brown paint, paint wood balls.

5. Spray frame and wood balls with spray sealer. Let dry. Distress wood balls using sandpaper.

6. With awl, make hole in top of two 1¾" wood balls and in bottom of 2" wood ball. Insert eye screws in holes, tightening with pliers. Make another angled hole in back of 2" wood ball.

7. Fill hole with glue, insert eye screw, and tighten. This will be center basketball. Back eye screw will be used for hanging.

8. With awl, make hole 1¼" from both sides on top edge of frame. Insert eye screws into these holes and tighten.

9. Cut chain in 6" lengths. Secure one side ball onto end of one chain. Pull open last link of chain with pliers and attach to eye screw in wood ball. Repeat for other side. Close open links.

10. Secure two lengths of chain to bottom of center wood ball. Bring one end of chain to right side and secure. Bring other end of chain to left side and secure.

11. Complete by **Securing Mirror** into frame.

Metallic Chain & Ball

Metallic Chain & Ball

Materials

- Acrylic gesso
- Chain: small links, silver (2')
- Eye screws: 15/32", silver (6)
- Mirror: cut to fit frame
- Spray paint: charcoal
- Wood balls: 1¼" (3)
- Wood frame: basic 11" square

Tools & Supplies

- Awl
- Glue: industrial-strength
- Pliers
- Sandpaper: fine grit
- Tack cloth
- Wire cutters

Instructions

Refer to General Instructions on pages 6-13.

1. **Make a Basic Frame** or use purchased frame.

2. Sand frame until smooth. Wipe with tack cloth. Seal with acrylic gesso.

3. Spray two coats of paint over entire frame, including inside lip. Let dry between coats. Spray wood balls. Let dry, rotate, and spray again.

4. With awl, make hole in top of two wood balls and in bottom of remaining wood ball. Insert eye screws into holes, tightening with pliers. Make another angled hole in back of ball with bottom hole.

5. Fill hole with glue, insert eye screw, and tighten. This will be center ball and back eye screw will be used for hanging.

6. With awl, make hole 1¼" from both sides on top edge of frame. Insert eye screws into these holes and tighten.

7. Cut chain into one 5", one 7", and two 6" lengths. Secure one side ball onto end of one chain. Pull open last link of chain with pliers and attach to eye screw in wood ball. Repeat for other side. Close open links.

8. Secure two lengths of chain to bottom of center ball. Bring one end of chain to right side and secure. Bring other end of chain to left side and secure.

9. Complete by **Securing Mirror** into frame.

Oak Vanity

Photo on page 21.

Materials

- Brass hinges: 1" (4)
- Knobs: oak, brass (6)
- Mirrors: 13" x 17" (1), 9" x 14 ½" (2)
- Oak board: 4" x 15" x ¼"
- Oak colonial stop: 1¼" x 15" x ⅜"
- Scrolled wood carving: 10½" x 2½"
- Spray sealer: gloss
- Wood stain: med. walnut

Tools & Supplies

- Drill and 9/64" bit
- Framing clamp
- Glue: wood
- Pencil
- Rags
- Sandpaper: fine grit
- Saws: miter, scroll saw
- Screwdriver
- Tracing paper
- Wire cutters

Instructions

Refer to General Instructions on pages 6-13.

1. Using miter saw and cutting on 45 degree angle, cut lengths from colonial stop: two 18", two 14", four 15",

and four 10". Lightly sand wood.

2. Place tracing paper over curved Pattern A below. Trace and cut out pattern. Trace pattern onto oak board. Using scroll saw, cut out pattern. Lightly sand wood.

3. Using clean rag, apply stain to all wood pieces.

4. Following Diagram A, glue frames together with wood glue and clamp until set.

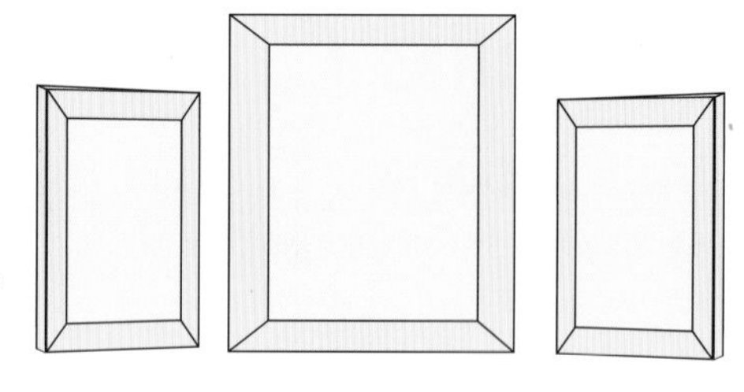

Diagram A

5. Center wood carving on curved oak piece and glue in place.

6. Glue curved piece to top edge of large frame, overlapping onto back about ½".

7. Spray frames with sealer.

8. Secure mirrors into frames.

9. Attach two hinges 1" from each end on both side mirrors. Attach frames together with hinges.

10. Drill hole on each side of bottom edge of each frame about 1¼" in from edges for mirror feet. Using wire cutters, cut screws included with knobs to ⅝" long. Twist cut end into knobs and twist other end into drilled holes in bottom of frames.

Pattern A Actual size

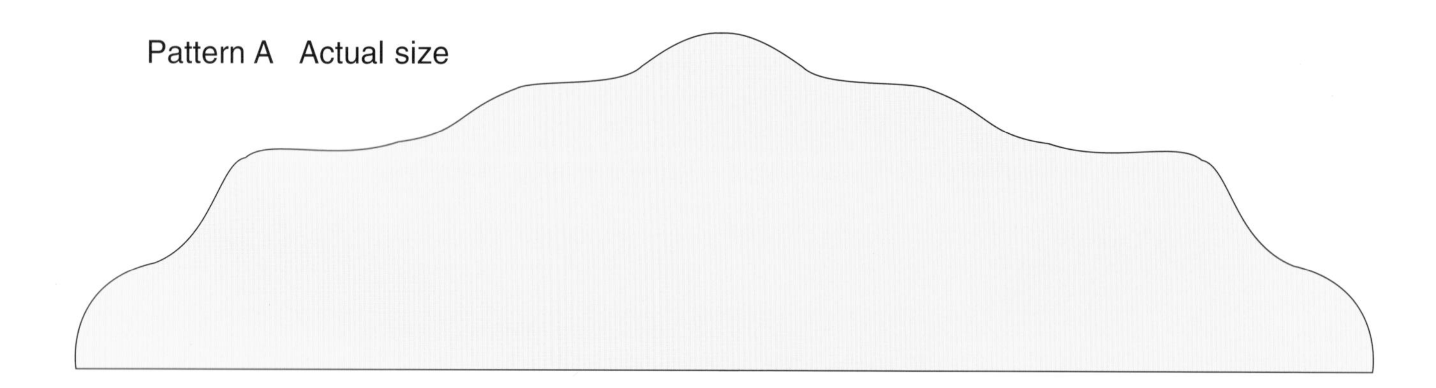

Edith Wharton

There are two ways of spreading light; to be the candle or the mirror that reflects it.

Oak Vanity

Pearls & Lace

Pearls & Lace

Materials

- Braid: ⅜", variegated metallic (1⅛ yd.)
- Brooch or large charm
- Gloves: cream with lace
- Mirror: round, 11"
- Pearl beads: 10mm (90), 7mm (130), 2mm pre-strung (1⅛ yds.)
- Silver beads: 7mm (120)
- Stamping glaze: gold
- Wood ring: flat, 12" x 1" x ¼", 10"-dia. center hole

Tools & Supplies

- Glue: industrial-strength
- Paintbrush: stiff
- Pencil

Instructions

Refer to General Instructions on pages 6-13.

1. Mark wood ring into quarters with pencil. Spread glue on one quarter of circle. Start on inside edge with 10mm pearl beads and make one row.

2. Repeat for second row with silver beads and third row with 7mm pearl beads. Continue in this manner for second, third, and fourth quarters until frame is covered.

3. Glue 2mm pearl beads around outside edge. Glue braid around side edge.

4. Paint over pearl surface with gold stamping glaze. Let dry.

5. Glue gloves and brooch in desired position.

6. Complete by **Securing Mirror** into frame.

Around the World

Photo on page 24.

Materials

- Acrylic paints: dk. blue, very dk. blue
- Color copies of Old World maps: large pages (6)
- Mirror: cut to fit smaller frame
- Webbing spray: verdi gris
- Wood frames: flat, larger frame with 2"-wide surface; smaller frame with 3"-wide surface (see photo)

Tools & Supplies

- Craft knife
- Glue: découpage, industrial-strength
- Paintbrush: ½"-wide, flat
- Newspaper

Instructions

Refer to General Instructions on pages 6-13.

1. Obtain color copies of maps.

2. Cut maps into various shapes and sizes with craft knife. **Découpage** maps over entire front of smaller frame, wrapping around inside and outside edges. Apply 3 coats of découpage glue, drying between coats.

3. **Secure Mirro**r into smaller frame.

4. Paint larger frame with very dk. blue. Let dry. Place frame in center of newspaper and spray with webbing. Let dry.

5. Thin dk. blue paint. **Wash** entire frame.

6. Apply industrial-strength glue around back edges of map mirror. Center mirror over outer frame, applying weight until glue sets.

Around the World

Metal Moose

Metal Moose

Photo on page 25

Materials

- Metal coat rack with design
- Mirror: cut to fit frame
- Spray sealer: matte
- Wood frame: ¾"-thick (frame size determined by coat rack size)
- Wood stain

Tools & Supplies

- Drill and screwdriver bit
- Propane torch
- Table saw
- Wood screws (2)

Instructions

Refer to General Instructions on pages 6-13.

1. **Make a Basic Frame** or use purchased frame.

2. Using propane torch, randomly burn spots on frame. Read and follow manufacturer's instructions for propane torch.

3. Stain frame. Let dry.

4. Spray frame with sealer.

5. Mark and drill holes to attach coat rack to bottom of frame. Screw rack onto frame with wood screws.

6. Complete by **Securing Mirror** into frame.

Window Screen

Materials

- Acrylic paint: dk. green
- Mirror: rectangle, ½" bigger than opening of screen
- Wooden window frame with screen

Tools & Supplies

- Gloves
- Masking tape
- Sandpaper: med. grit
- Wire cutters

Instructions

Refer to General Instructions on pages 6-13.

1. Using masking tape, cover front and back edges of screen.

2. Using dk. green paint, **Basecoat** entire frame being careful not to get paint on screen. Let dry.

3. Distress frame using sandpaper. *For more worn look, leave window screen out in weather for a week.*

4. Wearing gloves, cut and tear screen as desired. See photo on facing page.

5. Complete by **Securing Mirror** inside of frame of screen.

*Additional Ideas: The window screen can be embellished with silk or dried flowers. Use stenciling or other **Painting Techniques** on frame. Refer to General Instructions on page 8.*

Window Screen

Square Tiles

Square Tiles

Materials

- Acrylic gesso
- Acrylic paint: med. blue, lt. green, dk. terra-cotta, lt. terra-cotta, med. yellow
- Mirror: 10" x 15"
- Spray sealer: matte
- Wood board: ¾" x 3" x 5'
- Wood squares, ¼"-thick: 1" squares (40); 2" squares (16)

Tools & Supplies

- Glue: industrial-strength; wood
- Jigsaw
- Paintbrushes: ½"-wide flat; 1"-wide, flat

Instructions

Refer to General Instructions on pages 6-13.

1. From wood board, cut two pieces 21" long and two pieces 9" long with jigsaw. Using wood glue, attach pieces together to form frame as in Diagram A.

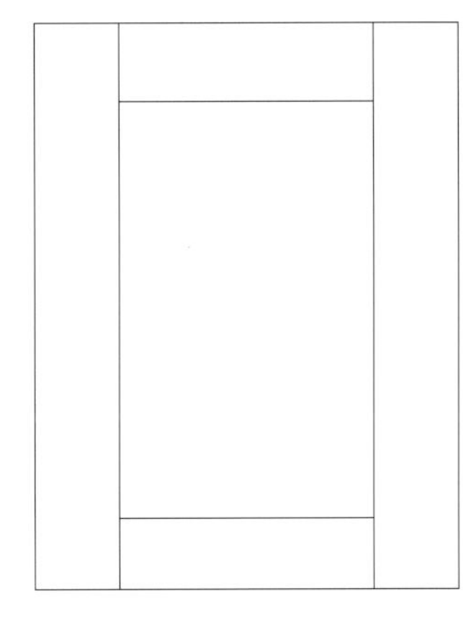

Diagram A

2. Seal with acrylic gesso.

3. Using both lt. and dk. terra cotta and ½" paintbrush, paint frame creating streaky background. Paint four 1" squares in this same manner.

4. Paint twenty 1" squares med. blue, streaking with a little lt. green and med. yellow.

5. Paint sixteen 1" squares med. yellow, streaking with a little lt. terra cotta and med. blue.

6. Using 1" paintbrush, paint 2" squares diagonally with med. yellow and lt. green.

7. Using industrial-strength glue, attach squares to frame following Diagram B.

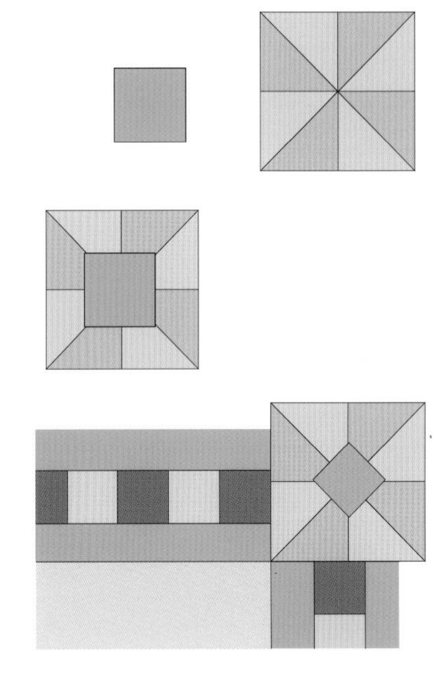

Diagram B

8. Spray frame with sealer.

9. Complete by **Securing Mirror** into frame.

Bird's Nest

Bird's Nest

Materials
- Acrylic gesso
- Acrylic paints: tan, white
- Bird's nest material: hay, green moss, Spanish moss
- Crackle medium
- Mirror: cut to fit frame
- Modeling paste
- Spray sealer: matte
- Twisted natural floral material
- Wood eggs: small (3)
- Wood frame: basic, 11" square

Tools & Supplies
- Cake decorating bag, coupler, #10 tip
- Glue: tacky
- Paintbrushes: 1"-wide, flat

Instructions
Refer to General Instructions on pages 6-13.

1. Make a Basic Frame or use purchased frame.

2. Seal frame with acrylic gesso.

3. Basecoat frame front and sides tan. Let dry.

4. Using flat paintbrush, apply **Crackle Medium** following manufacturer's instructions. Brush over crackle finish with white paint. Let dry.

5. With modeling paste in cake decorating bag, mold bird's nest onto right side of frame using #10 tip.

6. Cut small pieces of moss and hay. Press carefully into paste.

7. Paint eggs white. Speckle eggs with tan paint. Let dry. Glue eggs in nest.

8. Cut pieces of twisted material to fit around outer edges of frame. Glue in place. Cut several straighter pieces to use for tree branch. Glue in place.

9. Spray frame with sealer.

10. Complete by **Securing Mirror** into frame.

Garden Wall

Photo on page 33.

Materials
- Acrylic paint: black, med. brown, dk. green, med. green, lt. orange, med. pink, lt. purple, lt. yellow, white
- Craft sticks: small (11)
- Mirror: cut to fit frame
- Sculpting clay: white (4 oz.)
- Spray paint: stone texture
- Spray sealer: matte
- Texturizing medium (4 oz.)
- Wood frame: basic, 10" square

Tools & Supplies
- Bottle: small
- Glue: industrial-strength; tacky
- Paintbrushes: assorted
- Paper towels
- Toothpicks
- Wax paper

Instructions
Refer to General Instructions on pages 6-13.

1. Make a Basic Frame or use purchased frame.

2. With craft stick, thickly spread texturizing medium over entire front of frame, except bottom portion. Use craft stick to make lines in texturizing medium to look like a stone wall. Let dry overnight.

3. Spray entire frame with stone texture paint.

4. Wash grooves with thinned black paint. **Wash**

stones with thinned med. brown paint, blotting most of the paint off with paper towel. **Wash** again with thinned med. pink paint, blotting most of paint off with paper towel.

5. Cut seven craft sticks in half. Glue seven halves together in a fence fashion using a lot of tacky glue.

6. Wrap some wax paper around small bottle. Form sticks around bottle to create curve. Glue. Repeat with other seven halves on other side of bottle. Let dry.

7. Using **Sculpting Clay**, form two 1½" circles. Flatten backs of circles. Form two 1" rolled stems with leaves and one 1½" rolled stem with leaves.

8. For tulips, form about twenty-five ½" egg-shaped balls and press cross in top of each with toothpick.

9. For pansies, form about seventy-five ½" flat petals and twenty-five ¼" balls. Hook three petals together and add ball in center. Fold petals up slightly.

10. Bake clay according to manufacturer's instructions.

11. Carefully remove craft sticks from around bottle. **Wash** all sticks with thinned med. brown paint, including two remaining full-size sticks.

12. Paint flat clay circles med. green. Paint stems and leaves med. green. Shade stems and leaves with dk. green and highlight with white.

13. Paint tulips lt. yellow. Shade with lt. orange.

14. Paint top two petals of pansy lt. purple. Paint bottom petal lt. yellow. Add touch of lt. purple to inside of yellow petal. Paint pansy centers with mixture of med. brown and lt. yellow.

15. With industrial-strength glue, attach green circles to tops of uncut craft sticks.

16. Glue fence pots to each side of frame bottom.

17. Glue three tulips with stems and leaves across bottom of frame. Add a few pansies as desired.

18. Glue remaining pansies on one of the green circles. Glue topiary into one pot.

19. Glue remaining tulips on one of the green circles. Glue topiary into one pot.

20. Spray frame with sealer.

21. Complete by **Securing Mirror** into frame.

William Thackeray

The world is a looking-glass, and gives back to every man the reflection of his own face. Frown at it, and it in turn will look sourly upon you; laugh at it and with it, and it is a jolly kind companion.

Garden Wall

Lighthouse

Lighthouse

Materials

- Acrylic paints: black, blue, med. brown, gray, red, yellow, white
- Craft sticks: ¾" x 6" (5), ⅜" x 4½" (7)
- Mirror: to fit frame
- Sculpting clay: white (2 oz.)
- Spray sealer: matte
- Wood frame: basic 10" square with 3" surface

Tools & Supplies

- Aluminum foil
- Craft knife
- Glue: industrial-strength; wood
- Paintbrushes: assorted
- Pencil
- Ruler
- Sandpaper: fine grit
- Wire cutters

Instructions

Refer to General Instructions on pages 6-13.

1. Make a Basic Frame or use purchased frame. If needed, sand frame until smooth.

2. Following Diagram A, mark with pencil and draw general line for cliff, sky, and water. Mark lighthouse placement.

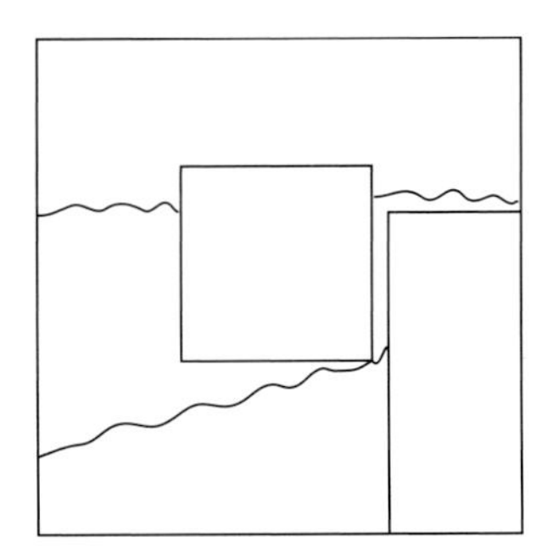

Diagram A

3. Dip craft stick into texturizing medium and fill in cliff area, making thicker in some areas. Let dry overnight.

4. Trace Pattern A from page 36. **Transfer** patterns onto large craft sticks. Cut out patterns with craft knife. Use wire cutters to snip off bottom and top. Sand edges. With pencil, number each piece according to pattern.

5. Following Diagram B, place stick #1 onto flat surface. Place sticks #2 on top and slide down to edge and glue. Repeat with sticks #3. Glue two small craft sticks to either edge, to lift it off of frame. Let dry. Set aside.

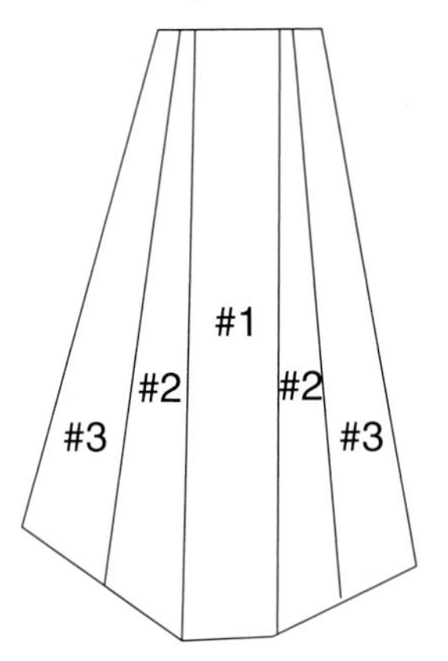

Diagram B

6. Using **Sculpting Clay,** form landing similar to Diagram C.

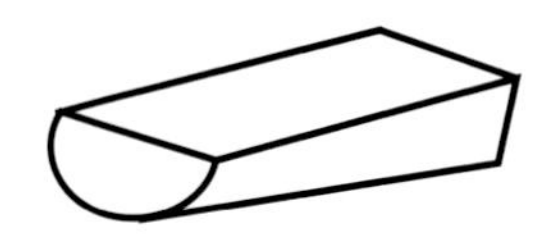

Diagram C

7. From small craft sticks, cut 4 sections each 1¼" long. Glue sections together as in Diagram D on page 36 to form windows. Let dry. Press windows onto clay landing, referring to photo. Press landing onto lighthouse base to form imprints. This will aid in construction once clay has hardened.

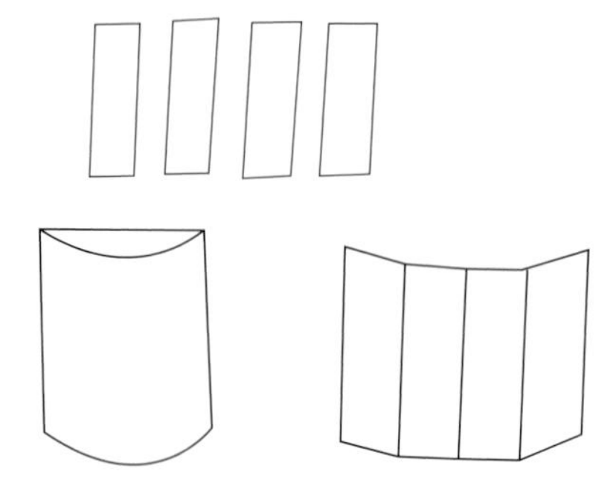

Diagram D

8. Sculpt roof and trim, referring to photo. Press roof and trim on windows to form imprints.

9. Remove clay pieces from lighthouse and bake according to manufacturer's instructions.

10. Paint lighthouse base red and white. Paint roof red. Paint trim white. Paint windows yellow and paint landing black. Let dry.

11. Assemble lighthouse with wood glue. Thin small amount of gray paint with water and wash over entire lighthouse.

12. On frame, paint sky and water blue. Mark horizon line.

13. For cliffs, pick up black, med. brown, and white on brush and dab onto cliffs. Blend going over entire cliff area. *Note: Carry painted areas around frame opening and onto back as this edge will reflect in mirror.*

14. For water, pick up blue, white, and black on brush and **Swirl** onto frame up to horizon line. The more paint used, the more texture water will have.

15. For sky, pick up blue and white on brush and stroke in back and forth motion. This will create shades and clouds. Let frame dry completely.

16. Glue lighthouse onto frame. Let dry.

17. Spray frame with matte sealer.

18. Complete by **Securing Mirror** into frame.

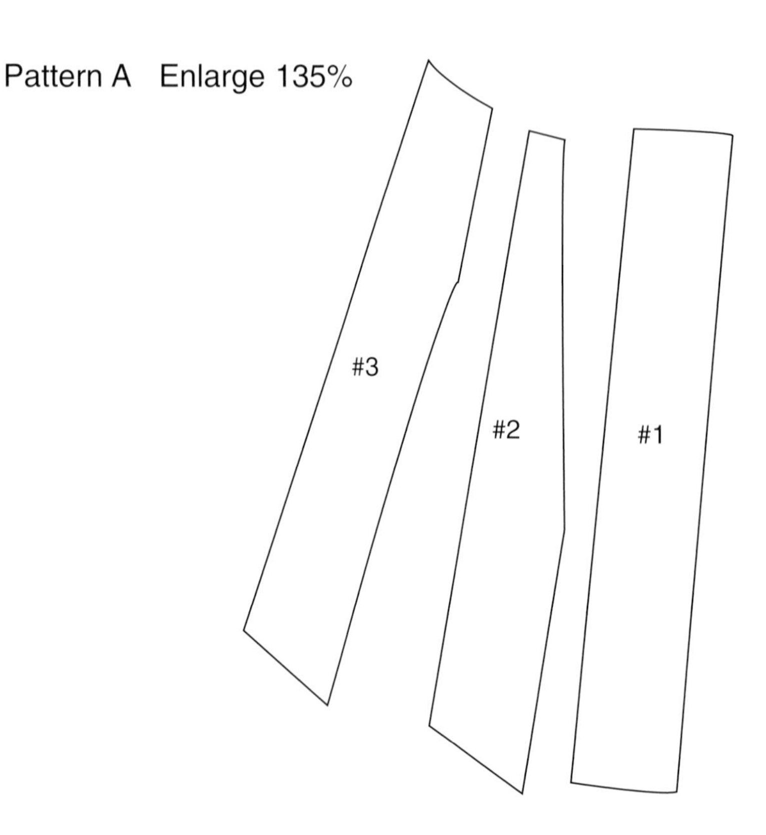

Pattern A Enlarge 135%

Seashore

Seashore

Photo on page 37.

Materials

- Acrylic paint: dk. brown, lt. brown, med. brown,
- Mirror: cut to fit frame
- Sculpting clay: white (2 oz.)
- Spray sealer: matte
- Spray paint: beige, stone texture
- Texturizing medium
- Wood frame: basic, 10" square

Tools & Supplies

- Glue: industrial-strength
- Knife: plastic
- Paintbrush: 1"-wide, flat
- Toothpicks

Instructions

Refer to General Instructions on pages 6-13.

1. Make a Basic Frame or use purchased frame.

2. With plastic knife, spread texturizing medium over front of frame, thicker in some parts. Let set until hard.

3. Spray front and sides of frame with stone texture paint.

4. Thin lt. brown paint with a little water and **Basecoat** entire frame. Paint over receded areas with med. brown paint. Using dk. brown paint, **Wash** edge of receded areas.

5. Using **Sculpting Clay**, roll out two snakes, one 8" long and one 7" long. With toothpick, draw lines in clay and poke out various areas to resemble twigs. Bake clay following manufacturer's instructions.

6. Wash over clay twigs with med. brown. Paint over twigs with dk. brown, making certain dark color gets into crevices.

7. Glue twigs onto frame as desired.

8. Spray entire frame with sealer.

9. Complete by **Securing Mirror** into frame.

Home Tweet Home

Photo on page 41.

Materials

- Acrylic paints: blue, brown, green, white, yellow
- Craft sticks: ¾" x 6" (7), ⅜" x 4½" (22)
- Mirror: 5½" square
- Sculpting clay: white (2 oz)
- Spray sealer: matte
- Texturizing medium
- Wood frame: basic, 10" square

Tools & Supplies

- Acrylic gesso
- Aluminum foil
- Craft knife
- Glue: industrial-strength
- Paintbrushes: assorted
- Pencil
- Ruler
- Sandpaper: fine grit
- Toothpicks

Instructions

Refer to General Instructions on pages 6-13.

1. Split three large craft sticks in half lengthwise. Mark, cut, and number as follows: three at 5¼" long, number 1, 2, 3; two at 4¾" long, number 4 and 5; one at 3" long, number 6. Sand edges of all craft sticks.

2. From small craft sticks cut 13 pickets for fence. Snip off bottoms and angle top. Sand edges. See Diagram A.

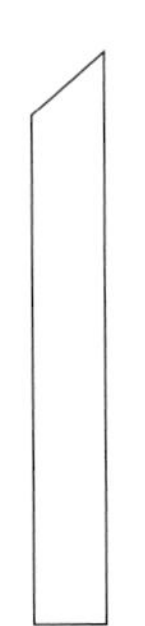

Diagram A

3. Place sticks #4 and #1 end to end for fence rails. Place sticks #5 and #2 end to end to make second fence rail. At joint, place one picket from step #2 on top. Glue in place. Let dry. See Diagram B on page 40.

4. Place remaining pickets from step #2 and stick #3 for birdhouse. Glue in place. Let dry. See Diagram B.

5. Cut large craft sticks into shingle sections: two 2"; two 1¾"; and two 1½". Stack and glue pieces together. See Diagram C.

Diagram C

6. Cut chimney pieces from small craft stick following Pattern A.

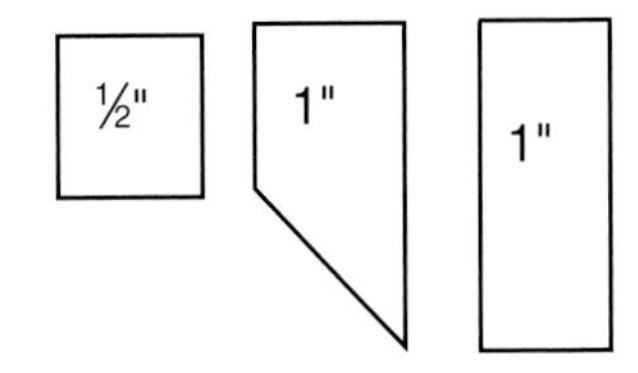

Pattern A

7. Cut small craft sticks for second set of shingles as follows: six 1" long, one split in half; three ¾" long. Sand edges.

8. Place three 1" sticks side by side on aluminum foil. Glue. Let dry. Repeat with three ¾" sticks. See Diagram D.

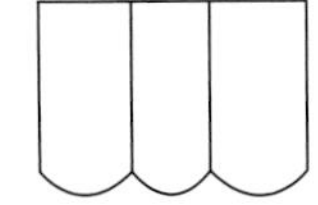

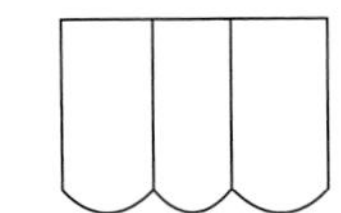

Diagram D

9. Place two 1" sticks in between two half sticks on aluminum foil. Glue. Let dry. See Diagram E.

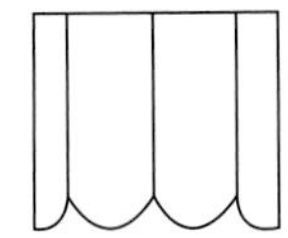

Diagram E

10. Stack grouped sticks from above to look like shingles. See Diagram F. Glue. Let dry.

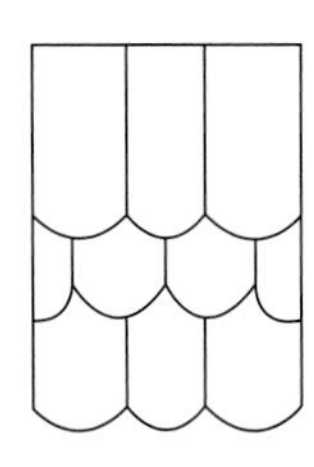

Diagram F

11. Using **Sculpting Clay**, form two birdhouses following Pattern B on page 40.

12. Using marble size of sculpting clay, pinch clay to make bird shapes. Form beaks by pinching clay. Form two wings for each bird in same manner. Use pieces of toothpick for bird's perch. Paint perches as desired.

13. Seal frame with acryic gesso.

14. On frame draw general line for grass and sky. Use blue paint to basecoat sky. Use green paint to **Basecoat** grass.

15. Using texturizing medium, paint over grass area again adding texture as desired.

16. Using white and blue paint, load paintbrush with both colors. Mix and blend as desired. Use back and forth stroke to shade sky. Let frame dry for two hours.

17. Paint fence and stick #6 white. Be certain to paint back of fence and along top edge.

18. Paint square-shaped birdhouse yellow and shade with white. Paint round-shaped birdhouse and chimney green mixed in white. Paint roofs brown. Paint birds brown shaded with blue. Paint their beaks yellow.

19. Glue fence in place as shown in photo on page 41. At top of stick #3, glue stick #6. This will serve as bird-house post. See Diagram B.

20. Glue roofs and chimney on birdhouses as shown in photograph. Glue birds where desired.

21. Spray frame with sealer.

22. Complete by **Securing Mirror** into frame.

Pattern B

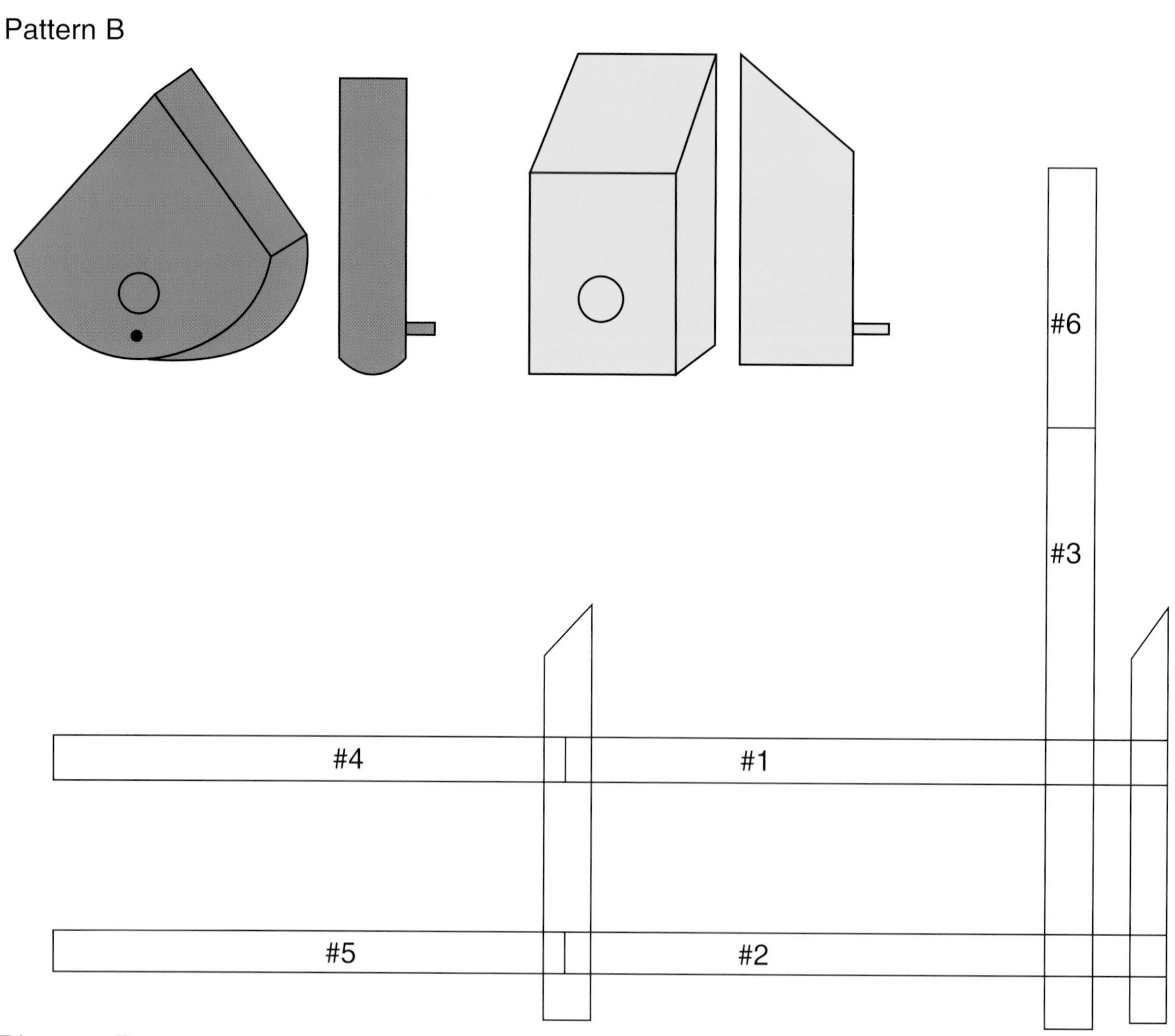

Diagram B

Home Tweet Home

Roses & Vines

Roses & Vines

Materials

- Acrylic gesso
- Acrylic paint: lt. pink, metallic copper, sage green
- Mirror: cut to fit frame
- Modeling paste
- Spray sealer: matte
- Wood frame: basic, 10" square

Tools & Supplies

- Cake decorating bag: coupler, pronged, #10, assorted tips
- Paintbrush: 1"-wide, flat
- Sandpaper: fine grit

Instructions

Refer to General Instructions on pages 6-13.

1. **Make a Basic Frame** or use purchased frame.

2. Seal with acrylic gesso.

3. **Basecoat** entire frame metallic copper. Lightly sand frame with sandpaper. Paint over copper paint with sage green.

4. Fill cake decorating bag with modeling paste. Make random vine pattern around frame surface. Switch to pronged tip and drag across vine to create texture.

5. Make random roses and buds along vine. Add leaves to roses and vines. Let dry for several hours or overnight.

6. Paint over vines and roses with sage green. Paint over roses and buds with lt. pink.

7. Lightly sand frame edges until copper shows through.

8. Spray frame with sealer.

9. Complete by **Securing Mirror** into frame.

Teacher's Busy

Photo on page 45.

Materials

- Acrylic paints: black, brown, metallic gold, med. green, off-white, dk. red, med. yellow
- Chain: brass, 4" length
- Eye hooks: brass (4)
- Jute: 3 ply (21")
- Mirror: 13" x 10½"
- Pencils: fat (3)
- Plaque: wood, 10" x 2½"
- Rulers: wood, 12" length (4)
- Spray sealer: matte
- Wood accessories: small, a, b, and c letters; apples (7); bell; book; school house; star
- Wood stain: med. walnut

Tools & Supplies

- Craft knife
- Drill and ⅜" drill bit
- Frame clamp
- Glue: industrial-strength; wood
- Graphite paper: white
- Paintbrushes: liner; ½"-wide, flat
- Pliers
- Rags
- Tracing paper
- Wire cutters

Instructions

Refer to General Instructions on pages 6-13.

1. Stain four rulers and wood plaque. Let dry. Spray with matte sealer.

2. Using wood glue, glue four rulers together to form frame. See Diagram A. Clamp at corners until set.

Diagram A

3. Paint center of wood plaque black. See photo on page 45.

4. Paint a, b, and c med. green, med. yellow, and dk. red.

5. Paint apples dk. red with brown stems and med. green leaves.

6. Paint bell metallic gold with brown handle; school house dk. red, brown, med. green, and med. yellow; star metallic gold; and book med. green with off-white pages and brown writing.

7. Using sharp craft knife, cut pencils so they are 1¾" long from eraser ends. Paint med. yellow.

8. Trace lettering and numbers from Pattern A. Using graphite paper, **Transfer** lettering to plaque and numbers onto six of the apples (two 1's, two 2's, two 3's).

9. Using liner brush, paint lettering and numbers off-white.

10. Spray all wood pieces with sealer.

11. Complete by **Securing Mirror** to back of ruler frame.

12. Referring to photo, use industrial-strength glue to attach a, b, and c, bell, book, school house, and star. Glue one apple at top. Glue pencils evenly across bottom edge of frame.

13. Drill hole in top center of each numbered apple. Cut jute in three equal pieces. Glue one end of jute into one #1 apple and other end into other #1 apple. Repeat with other apples.

14. Tie loop in center of each strand. Use loop to hang on pencil "hooks."

15. Twist two eye hooks into top of frame, 3" in from each side. Twist other two eye hooks into bottom edge of plaque 2" in from each side.

16. Using wire cutters, cut chain in half. Using pliers, attach chain to eye hooks on each side of plaque and hang ruler frame.

Pattern A Actual size

teacher's busy
take a number

Pattern B Actual size

1 2 3

Teacher's Busy

Chapter Two
Techniques on Mirrors

No Place Like Home

No Place Like Home

Materials

- Acrylic gesso
- Acrylic paints: black, navy blue, cream, red, white
- Balsa wood: 18" x 4" x ¼"
- Corner brackets: 1½" (4)
- Mirror: unframed, 24" square
- Spray sealer: satin
- Wood moulding: 30" x 3¼" x ¾" (4 pieces)
- Wood letters, 1¼"-high spelling: THERE IS NO PLACE LIKE HOME (twice)

Tools & Supplies

- Aluminum foil
- Bowl: paper
- Craft knife
- Glue: wood
- Framing clamp
- Paintbrushes: ¾"-wide, flat; thin, old
- Sandpaper: med. grit
- Saw: miter
- Screwdriver
- Wood putty

Instructions

Refer to General Instructions on pages 6-13.

1. Make a Basic Frame or use a purchased frame. Miter ends of molding to 45°angles. Place moulding face down on wood.

2. Carefully screw corner brackets to each corner.

3. Turn right side up. Fill any gaps in joints with wood putty.

4. Using craft knife, cut four pieces from balsa wood, each ½" x 18". Sand edges until smooth.

5. Lightly sand letters, if needed. Seal all wood with acrylic gesso.

6. Basecoat frame and four balsa strips white.

7. Thin cream paint with water and **Wash** over pieces. Lightly distress with sand-paper.

8. Loading paintbrush with both navy blue and black paint, paint **Checkerboard Stroke** around frame. *Note: The checks need not be even.* When dry, lightly distress with sandpaper.

9. In paper bowl, dilute red paint with small amount of water. Dip letters in paint until completely covered. Set on aluminum foil and let dry.

10. Spray frame and letters with sealer.

11. Complete by **Securing Mirror** into frame.

12. Using paintbrush to apply wood glue, attach balsa strips in center of mirror, forming square. See photograph on page 46.

13. Evenly space letters around mirror to spell "THERE IS NO PLACE LIKE HOME" twice. Glue letters onto mirror.

Moonrise

Moonrise

Materials

- Acrylic paints: black, 2 different bronzes, metallic gold
- Mirror: round, 13"
- Spray sealer: matte
- Tooling brass: 36 gauge (15" x 10")
- Wood circle: ¾"-thick (13½"-dia.)

Tools & Supplies

- Awl
- Cloth
- Clothes pins
- Contact paper
- Craft knife
- Etching cream
- Glue: industrial-strength
- Graphite paper
- Paintbrushes: assorted
- Paper towels
- Pen: ball point
- Router
- Sandpaper: fine grit
- Scissors: old

Instructions

Refer to General Instructions on pages 6-13.

1. Place mirror on top of contact paper and trace around it. Cut out circle and adhere it to front of mirror.

2. Enlarge Pattern B on page 50. **Transfer** pattern using graphite paper onto contact paper. Using sharp craft knife, cut away part of pattern that will be etched.

3. Paint etching cream onto mirror, covering well. Let set about 5 minutes, then wash off with water. Dry well and peel off contact paper.

4. Enlarge Pattern A on page 50. Cut out pattern and place on brass sheet. Lightly trace around pattern. Cut out shape from brass with old scissors. Place pattern back over brass and trace over details with pen. Indentations should show through onto brass.

5. Remove pattern and continue going over lines with pen and awl until desired effect is reached.

6. Paint over brass moon with metallic paints, blending and swirling to create desired effect.

7. Completely cover brass moon with black paint. Wipe off with paper towel to create an antique look. Remove excess paint from lines with awl.

8. Spray moon with sealer.

9. Glue moon on mirror and hold in place with clothes pins until dry.

10. Route out center of wood circle, leaving ¼" ledge. Place mirror into center to check fit. Remove mirror. *Note: If desired, use an already framed mirror and paint frame as in step 11.*

11. Sand wood circle. Paint sides and front ledge of mirror to match moon.

12. Spray with sealer.

13. Glue mirror in place. Place cloth over mirror and lay semi-heavy object on mirror until set.

Pattern A Enlarge 195%

Pattern B Enlarge 220%

Family Heirloom

Photo on page 51.

Materials

- Embellishments, antique: chains; necklace; old keys
- Eyebolt: extra small
- Family photographs
- Marker: permanent, extra fine-tip, gold
- Mirror: to fit frame
- Wood frame: 2¼"-thick surface, 8½" x 10½" opening

Tools & Supplies

- Cloth
- Craft knife
- Glue: découpage
- Graphite paper: gray
- Heavy paper/ cardboard: several sheets
- Scissors
- Spouncer brush
- Tracing paper
- Varnish: polyurethane semi-gloss, water-based

Family Heirloom

Instructions

Refer to General Instructions on pages 6-13.

1. Arrange family photographs on sheets of heavy paper or cardboard. Temporarily attach photos. Obtain color copies of each page. Cut out each photo from copies with scissors.

2. **Découpage** color copies onto frame varying sizes, shapes, and positions to create interest. Completely cover frame.

3. Using spouncer, brush on four coats of polyurethane varnish, letting dry completely between coats.

4. Using craft knife, make small hole in inside top edge about ¼" from right side. Twist small eyebolt into frame and attach embellishments as desired.

5. Trace lettering from Pattern A. Using graphite paper, **Transfer** lettering onto mirror. Go over lettering with gold marker. Let dry overnight. Clean carefully with damp cloth to remove graphite residue.

6. Complete by **Securing Mirror** into frame.

Pattern A Actual size

At times
I can almost
feel the presence
of my ancestors—
A gentle, guiding touch
from those who've gone before.

Gilded Window

Gilded Window

Photo on page 53.

Materials

- Acrylic paints: black, cream, mustard yellow, white
- Liquid leaf: brass
- Windowpane mirror: 16" square

Tools & Supplies

- Graphite paper: gray
- Paintbrushes: fine-tip liner, 1"-wide, flat
- Toothbrush: old

Instructions

Refer to General Instructions on pages 6-13.

1. Transfer Pattern A onto left side of mirror, using graphite paper. Use liner brush, paint lettering with liquid leaf.

2. Mask-off mirror portions to protect from paint. **Basecoat** frame white. Thin cream paint with water and **Wash** over frame.

3. Float edges with mustard yellow paint.

4. Thin black paint with a little water. Using liner brush, randomly paint words around outside of frame.

5. Using toothbrush, lightly **Spatter** black paint over frame. See photo on page 53.

Pattern A Actual size

With every rising of the sun,

Think of your life as just begun

Beveled Glass

Photo on page 55.

Materials

- Board: old, flat, larger than 16" square
- C-lead edging: 1 strip
- Charm: antique
- Copper patina for lead
- Copper tape
- Fingernail polish: clear
- Glass pieces: beveled 2" x 8" (4), 2" x 2" (4), 1" x 2" (16)
- Mirror: beveled, 12" square
- Nails
- Soldering flux
- Spray sealer: matte
- Wire solder: 60/40 grade

Tools & Supplies

- Glue: industrial-strength
- Hammer
- Masking tape
- Paintbrush: old
- Soldering iron
- Toothbrush: old

Beveled Glass

Instructions

Refer to General Instructions on pages 6-13.

1. Seal edge of mirror with clear fingernail polish. Do not get polish on front of mirror. Brush polish over onto back about ¾". Let dry.

2. Spray back of mirror with sealer.

3. Lay out mirror and place beveled glass pieces around mirror. Tape edges of glass and mirror with copper tape. Make certain tape is even on all sides as soldering adheres to it.

4. Place mirror with beveled glass sides on wooden board and hammer nails around edges to hold mirror in place.

5. Paint all exposed copper tape areas with flux. Do not get flux on skin. Follow manufacturer's instructions for precautions.

6. With **Soldering** iron, solder all seams. Let cool.

7. Remove nails from board. If desired, flux and solder back side of mirror.

8. Cut C-lead edging to fit edges. Push edge onto each side, holding in place with masking tape.

9. Flux corners and at each point where copper meets edging. Solder together. See Diagram A.

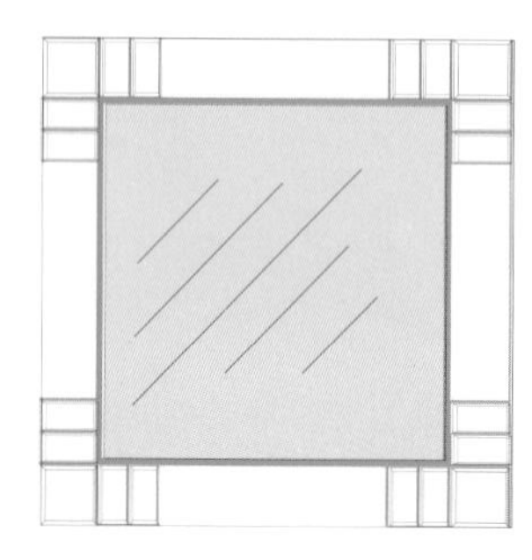

Diagram A

10. Clean glass by using toothbrush with soap and water.

11. Using old paintbrush, apply copper patina over all soldered areas until desired color is achieved.

12. Wash mirror with water.

13. Add charm to mirror with industrial-strength glue.

Scroll Saw Silhouette

Scroll Saw Silhouette

Photo on page 57.

Materials

- Acrylic gesso
- Acrylic paints: lt. blue, dk. green, lt. green, dk. pink, lt. pink, med. purple, dk. yellow, lt. yellow, white
- Mirror: cut to fit frame
- Spray sealer: matte
- Wood frame: with three relief surfaces on molding; 15" square
- Wood: hardwood, 12" x 12" x ¼"

Tools & Supplies

- Glue: wood
- Graphite paper
- Paintbrushes: fine-tipped liner; ½"-wide, flat
- Paper towels
- Sandpaper: fine grit
- Scroll saw
- Sponge
- Tracing paper

Instructions

Refer to General Instructions on pages 6-13.

1. Transfer Pattern A and B on page 59 onto wood using graphite paper. Cut out pattern with scroll saw. Lightly sand pieces until smooth.

2. Marbleize both sides of wood cut-outs by brushing on lt. yellow basecoat. Sponge on dk. yellow, dk. pink, and white. Veins are made with dk. yellow and dk. pink.

3. Thin lt. yellow paint with water and **Wash** over both pieces.

4. Seal frame with acrylic gesso.

5. Paint **Checkerboard Stroke** around frame with lt. and dk. yellow. See photo.

6. Paint swirled stripe design around frame with dk. pink, dk. green, and dk. yellow.

7. Paint stripe design around outer frame edge and inner lip with lt. blue and med. purple. See photo on page 57.

8. Thin white paint with water and **Wash** over stripe designs, but not checker-board. Wipe off slightly with paper towel to achieve milky look.

9. Randomly paint roses around checkerboard design by making abstract circles with lt. pink paint. Add detail to roses with dk. pink.

10. Use dk. green paint to add leaves, highlighting with lt. green.

11. Spray frame and silhouette pieces with sealer.

12. Complete by **Securing Mirror** into frame.

13. Attach silhouette pieces onto inside corners of frame with wood glue.

Pattern A Actual size

Pattern B Actual size

Floral Mirror

Floral Mirror

Materials

- Foam stamping sheet: approximately 9" x 8"
- Glazes: brown, burgundy, green, dk. green, red, silver, white, yellow
- Mirror: framed

Tools & Supplies

- Craft knife
- Paint brushes: ¼"-wide, flat; ½"-wide, flat
- Paper towels
- Pencil: white

Note: To make this mirror seasonal, stamp leaves and vines on frame and flowers on mirror only. Flowers can be washed off and a new pattern can be stamped for next season. Practice stamping on paper before proceeding on mirror.

Instructions

Refer to General Instructions on pages 6-13.

1. Transfer Pattern A on page 62 onto stamping sheet using white pencil. Cut out with craft knife.

2. Cut detail into design with craft knife following stamping sheet manufacturer's instructions.

3. Clean mirror of all dust and fingerprints.

4. Make pink mixture from red and white glazes and brush onto large petal stamp. Use long strokes. Add a little burgundy to brush and paint highlights onto stamp. Repeat with a little brown.

5. Firmly press stamp onto mirror in desired position. Continue stamping and overlapping until flower is completed. Do not reload brush between stamps. The first petal will be dark and following petals will be lighter.

6. Continue stamping flowers randomly, loading stamp for each flower. Each flower will be individual if colors are applied slightly different. Use both small and large petal stamps.

7. For leaves, load stamp with dk. green. Add green and silver as desired. Stamp leaves randomly around flowers, loading stamp when necessary.

8. Vines are painted freehand with long, squiggly lines-some dark and some light. Detail can be added to flower centers with brown and yellow as desired.

Pattern A Actual size

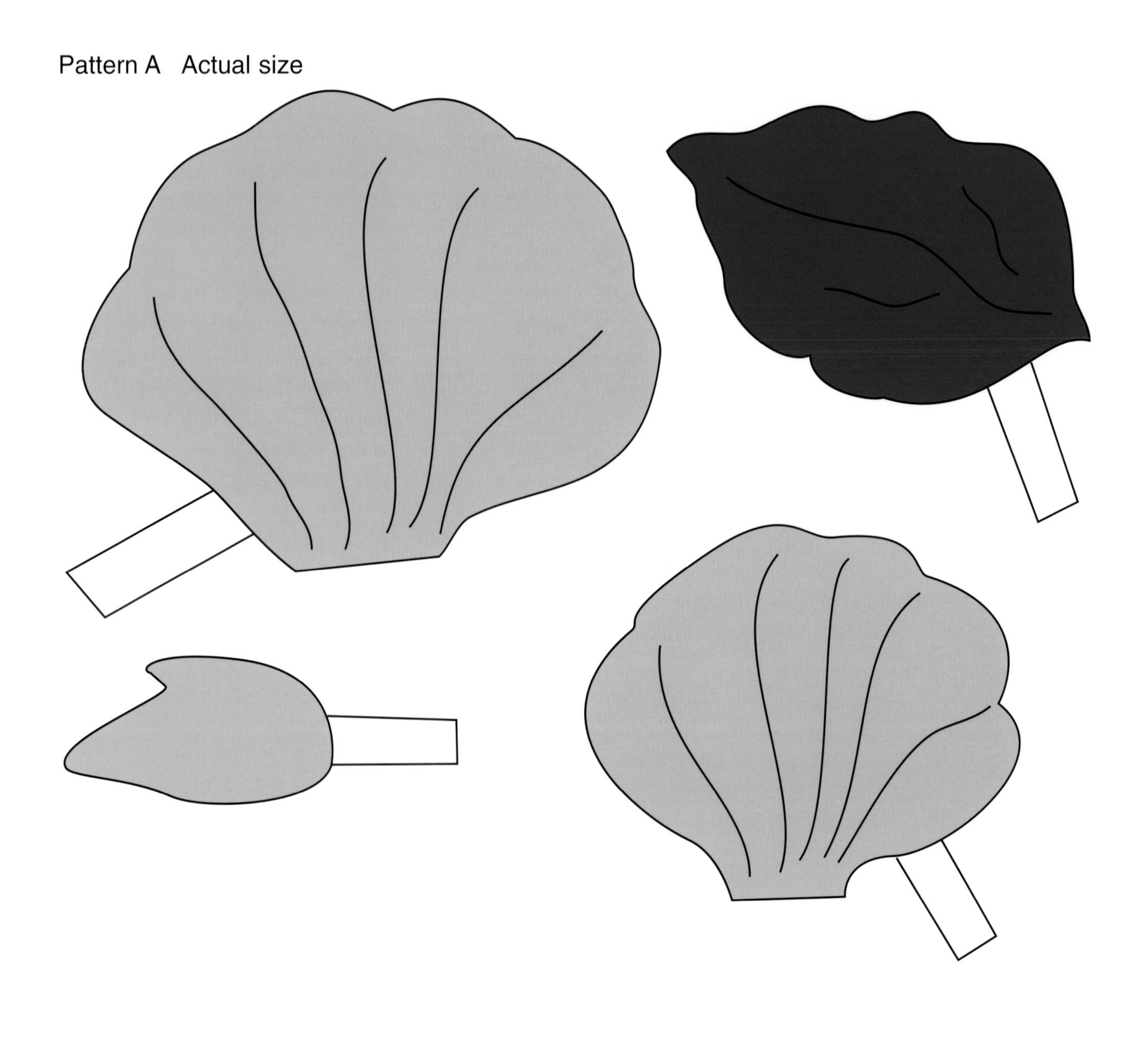

Lewis Carroll

"How would you like to live in Looking-glass House, Kitty? I wonder if they'd give you milk there? Perhaps Looking-glass milk isn't good to drink–but oh, Kitty! now we come to the passage. You can just see a little peep of the passage in Looking-glass House, if you leave the door of our drawing-room wide open: and it's very like our passage as far as you can see, only you know it may be quite different on beyond. Oh, Kitty, how nice it would be if we could get through into Looking-glass House! I'm sure it's got, oh! such beautiful things in it! Let's pretend there's a way of getting through into it, somehow, Kitty.

Sundial

Look toward the Sun
XII
III
IX
VI
And the shadows will fall behind you.

Sundial

Photo on page 63.

Materials

- Acrylic paint: golden brown, lt. brown, sage green, med. yellow, white
- Balsa wood: 10" x 1⅛" 1/16"
- Dowel ¼"-dia. (7" length)
- Mirror: round, cut to fit frame
- Stenciling sponge: small foam
- Wood frame: round with flat surface, 12"-dia. opening

Tools & Supplies

- Craft knife
- Dremel electric rotary tool with rubber tip attachment
- Glue: clear cement; wood
- Graphite paper: white
- Paintbrushes: fine-tip liner; ½"-wide, flat
- Tape: transparent
- Tracing paper

Instructions

Refer to General Instructions on pages 6-13.

1. Using flat brush, **Basecoat** frame medium yellow.

2. **Transfer** Pattern A from page 65 onto foam sponge. Cut out with craft knife. Stencil leaf shapes around frame with lt. brown, sage green, and white paint.

3. Thin sage green paint with water. **Wash** thinned paint around outside edge of frame. Repeat around top edge of frame with thinned lt. brown paint.

4. Enlarge Pattern B and Pattern C on page 65. Using graphite paper, **Transfer** lettering onto flat surface of frame. See photo on page 63.

5. Using liner brush, paint lettering lt. brown. Add shadows around lettering with golden brown paint.

6. Enlarge Pattern D on page 65. Using graphite paper, **Transfer** dial design onto balsa wood. Cut out with sharp craft knife.

7. Using wood glue, attach dowel across top edge of dial, extending ¼" past inside point of dial. Let set.

8. Paint dial med. yellow. Wash dowel with sage green. **Wash** inside and outside edges of dial with lt. brown.

9. Obtain color copies of Pattern E on page 65.

10. Draw intersecting lines across back of mirror to help with placement of numbers.

11. Place tracing paper over Pattern E on page 65 and draw general outline. Invert tracing paper and center over line at top of mirror, ½" in from edge. Tape tracing paper in place. Tuck graphite paper under design and trace outline. Repeat with other numbers. Remove designs.

12. Using rubber tip attachment at low speed on dremel tool, rub off silvering from back of mirror where numerals will be placed.

13. Cut out Roman numerals and tape in position on back of mirror.

14. Complete by **Securing Mirror** into frame.

15. Using clear cement, attach dial with dowel side up at desired time.

Pattern A Actual size

Pattern D Enlarge 180%

Pattern B Enlarge 120%

Pattern E Actual size

Pattern C Enlarge 145%

Chapter Three
Unique Mirror Ideas

Garden Gate

Garden Gate

Materials

- Acrylic paint: antique white
- Button
- Charms: to resemble hinges (2)
- Mirror: 6" x 7"
- Spray sealer: matte
- Wood plant markers: 12" x 1" (8)
- Wood strip: 17" x 1" x ⅜"

Tools & Supplies

- Glue: wood
- Hand saw
- Paintbrush: 1"-wide, flat
- Paper
- Ruler
- Sandpaper: fine grit

Instructions

Refer to General Instructions on pages 6-13.

1. Cut wood strip into two 8½" lengths. Draw two horizontal lines across paper 7" apart. Place wood strip across each line. Remove paper. See Diagram A.

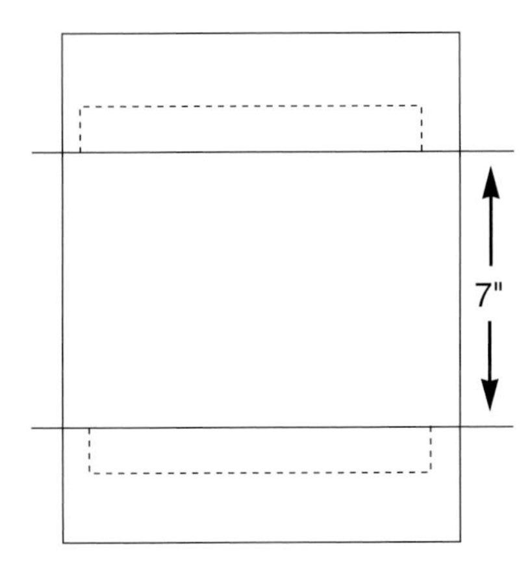

Diagram A

2. Place eight garden markers vertically across strips of wood in jagged manner. With wood glue, attach two outside stakes on each side. Remove four stakes in center. See Diagram B.

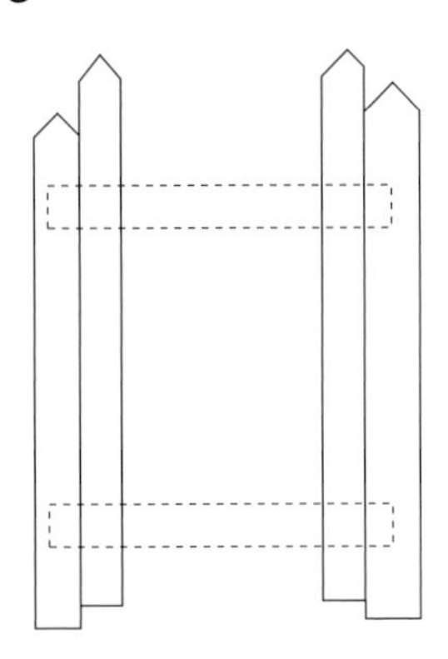

Diagram B

3. From center stakes, cut two 4" lengths and two 3¾" lengths from top pickets. Cut two 2½", one 2¼", and one 2" length from bottom of pickets.

4. Wood-glue center pickets in place referring to Diagram C.

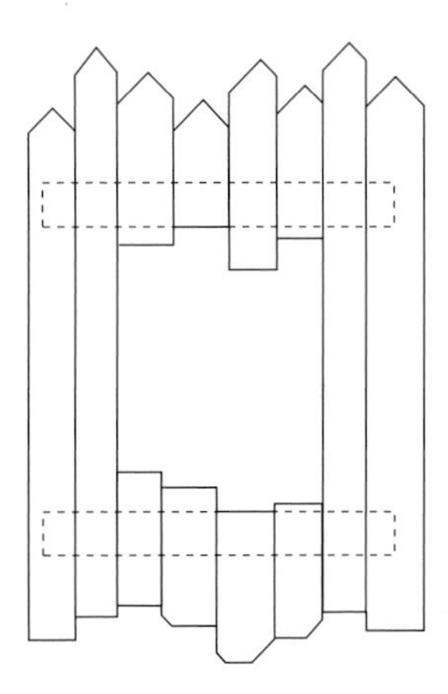

Diagram C

5. **Wash** pickets with thinned antique white paint.

6. Lightly sand over wood to create distressed look.

7. Spray wood with sealer.

8. Glue charms in place for hinges and button in place for handle.

9. Complete by **Securing Mirror** into frame.

Fly Fishing

Fly Fishing

Materials

- Acrylic gesso
- Acrylic paint: black, dk. brown
- Bead: wooden, oval, ⅝"
- Burlap (½ yd.)
- Cattails: decorative
- Cotton string: 24"
- Fishing fly
- Mirror: cut to fit frame
- Painted fish: about 4"
- Patina finish: green
- Self-adhesive copper: ¼"-wide (2 yds.)
- Spray sealer: matte
- Stem wire: 18 gauge (12")
- Tree branch
- Wood frame: flat, 15" x 13"
- Wood stain: med. walnut
- Wood wheel: 2"

Tools & Supplies

- Glue: industrial-strength; tacky
- Paintbrushes: ¼" -wide, flat, 1"-wide, flat; old
- Rag
- Ruler
- Tea bags
- Wire cutters

Instructions

Refer to General Instructions on pages 6-13.

1. Seal frame with acrylic gesso.

2. Basecoat entire frame dk. brown. Let dry. Spray with sealer.

3. Tea-dye cotton string.

4. Using rag, stain tree branch, wheel, and bead.

5. Place frame right side down onto burlap. Measure 2" from outside edge of frame and trim excess burlap. Measure 1" from inside edge of frame and cut out center. Miter cut corners.

6. Using old paintbrush, apply thin layer of tacky glue to front and sides of frame. Press and smooth burlap around frame. Wrap edges to back and glue in place.

7. Using wire cutters, cut 5" piece of stem wire. Make small loop at end of wire. Slide on bead and bend wire at 45° angle. Slide on wheel and wrap around one end of tree branch. Cut two 3" pieces of wire. Make two loops and attach one to center of branch and one to other end.

8. Complete by **Securing Mirror** into frame.

9. Cut strips of self-adhesive copper to fit around center of frame. Press in place.

10. Thin black paint with water. Carefully **Wash** over copper with thinned paint, blotting with rag continuously.

11. Carefully brush patina over copper. If decorative cattails have brass or copper accents, brush these with patina.

12. Using industrial-strength glue, attach cattails to left bottom corner of frame. Glue fish in place. Glue "pole" at top right corner and run cotton string through loops. See photo on page 68.

13. Secure fishing fly to end of string, then glue fly to mirror.

Checkerboard Mirror

Checkerboard Mirror

Materials

- Acrylic gesso
- Acrylic paints: black, dk. brown, lt. brown, med. brown
- Mirror: 13¾" x 4½" (2); 13¾" x 8½"
- Spray sealer: matte
- Tiles, square, 1": white (55); black (55)
- Wood: plywood, paint grade, 26" square
- Wood moulding: 25¾" x 4½" x ¾"

Tool & Supplies

- Aluminum foil
- Cloth
- Glue: industrial-strength; wood
- Paper towels
- Paintbrushes: assorted
- Sandpaper: med. grit

Instructions

Refer to General Instructions on pages 6-13.

1. From wood, cut one piece 7½" x 24", four pieces 2" x 13", and one piece 2¾" x 24" to make frame. Assemble frame using wood glue. See Diagram A on page 72.

2. Seal with acrylic gesso.

3. Using sandpaper, sand shelf until smooth. Wipe dust off with cloth.

4. Dip paintbrush in three different browns so paint is thick on brush. **Swirl** paint on top and middle pieces of frame. Continue until pieces are covered.

5. Take dk. brown and black paint and swirl in over previously painted brown areas giving them a marble look.

6. Take paper towel and crinkle it up, dip in water and squeeze, leaving some water in paper towel. Dip paper towel into lt. brown paint and dab onto aluminum foil to mix water and paint. Dab on frame creating diagonal lines.

7. Paint moulding and bottom of frame with black paint.

8. Attach moulding on top of frame using wood glue. See Diagram A on page 72.

9. Spray frame with sealer.

10. Measure placement of tiles and apply industrial-strength glue to this area. Let glue set while applying glue to backs of tiles. Let both set until tacky.

11. Apply tiles in checker-board manner. See photo on page 70.

12. Complete by **Securing Mirrors** into frame.

Diagram A

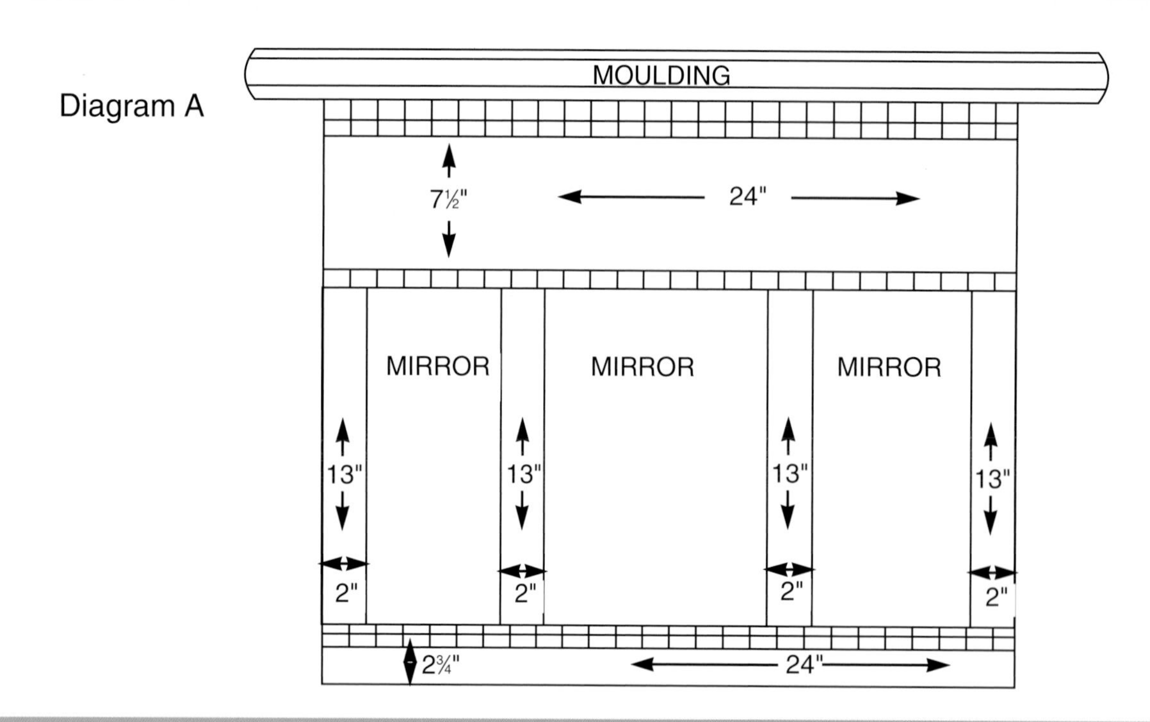

Spiked Mirror

Materials

- Bailing wire: 16 gauge (15')
- Brass ring : 6"-dia.
- Cardboard circle: 5⅞"-dia.
- Electrical tape: black
- Galvanized wire: 20 gauge,(15')
- Mirror: round, 6¼"
- Spray paint: black

Tools & Supplies

- Glue: industrial-strength
- Masking tape
- Razor blade
- Wire cutters

Instructions

1. Cut off 40" of 16 gauge wire and bend in half. Bend in half again. This will be mirror handle.

2. With remaining 16 gauge wire, twist one end around one end of handle and randomly shape around brass ring forming spikes at about 1" intervals. See Diagram A.

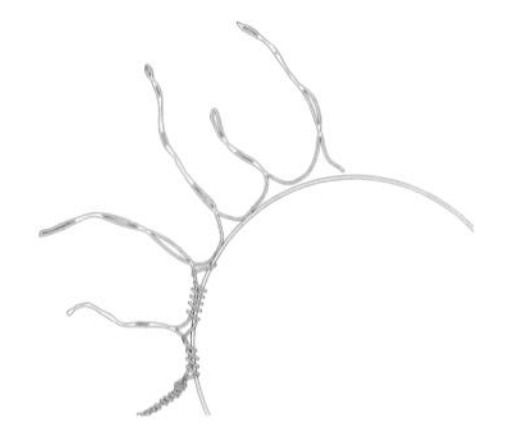

Diagram A

3. Wrap handle with electrical tape.

4. Using 20 gauge wire, wrap all around handle and brass ring.

5. Spray with black paint.

6. Spray back of mirror with black paint.

7. Glue mirror onto back of brass wire. Place heavy object onto mirror to hold firm until set.

8. Tape cardboard circle over mirror. Apply second coat of black paint covering all glued areas.

9. Remove cardboard from mirror. Using razor blade, scrape any excess paint from mirror.

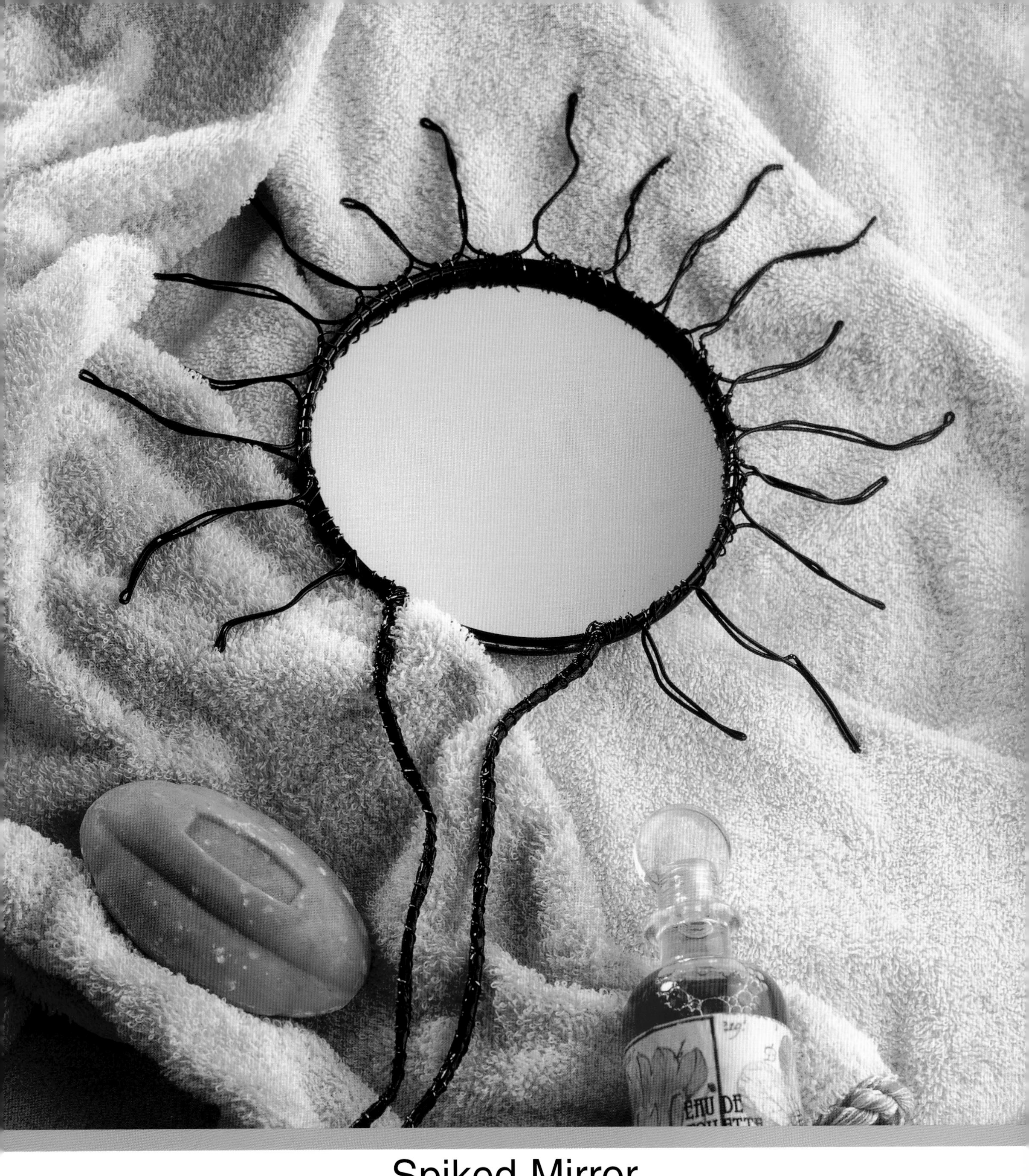

Spiked Mirror

Bunny Mirror

Bunny Mirror

Materials

- Acrylic gesso
- Acrylic paints: black, lt. ivory, periwinkle, lt. rose, tan, white
- Marker: permanent, fine-tip, black
- Mirror: oval, 8"
- Spray sealer: gloss; water based
- Wood: pine, 12" x 14" x ½"

Tools & Supplies

- Glue: industrial-strength
- Graphite paper: gray
- Paintbrushes: liner, ¼"-wide; 1"-wide spouncer
- Sandpaper: med. grit
- Scroll saw (or jigsaw)
- Toothbrush: old
- Tracing paper

Instructions

Refer to General Instructions on pages 6-13.

1. Enlarge and transfer Pattern A, B, C, and D on page 76.

2. Using graphite paper, **Transfer** designs on wood tracing outlines only.

3. Cut out wood design with scroll saw or jigsaw.

4. Sand wood until smooth. Seal by applying acrylic gesso. Allow to dry completely, then sand again lightly.

5. Using spouncer brush, **Basecoat** all pieces with lt. ivory paint.

6. Using graphite paper, **Transfer** all inside pattern lines on wood pieces.

7. Paint nose and eyes black. Paint foot pads and inner ear with lt. rose. Lightly **Stipple** lt. rose onto cheeks.

8. Thin white paint with a little water. Using liner brush, paint lines on paw pads and inner ear.

9. Stipple all outside edges with tan paint. Using ¼" paintbrush, shade all outside edges with tan paint.

10. Thin periwinkle paint with a little water. Using liner brush, paint squiggly lines on paw pads and inner ear.

11. Using ¼" paintbrush, shade white paint on top of cheeks and add highlight to nose.

12. Add freckles by dipping opposite end of paintbrush into black paint and dotting onto cheeks.

13. Thin tan paint with a little water. Dip toothbrush into paint and lightly splatter all pieces.

14. With fine-tipped black marker, outline all inside pattern lines with broken line stroke.

15. Spray with sealer.

16. Glue mirror to bunny body, using guide marks for placement.

17. Glue remaining pieces to bunny, using guide marks for placement. *Note: It may be necessary to clamp pieces together until they set to avoid slipping.*

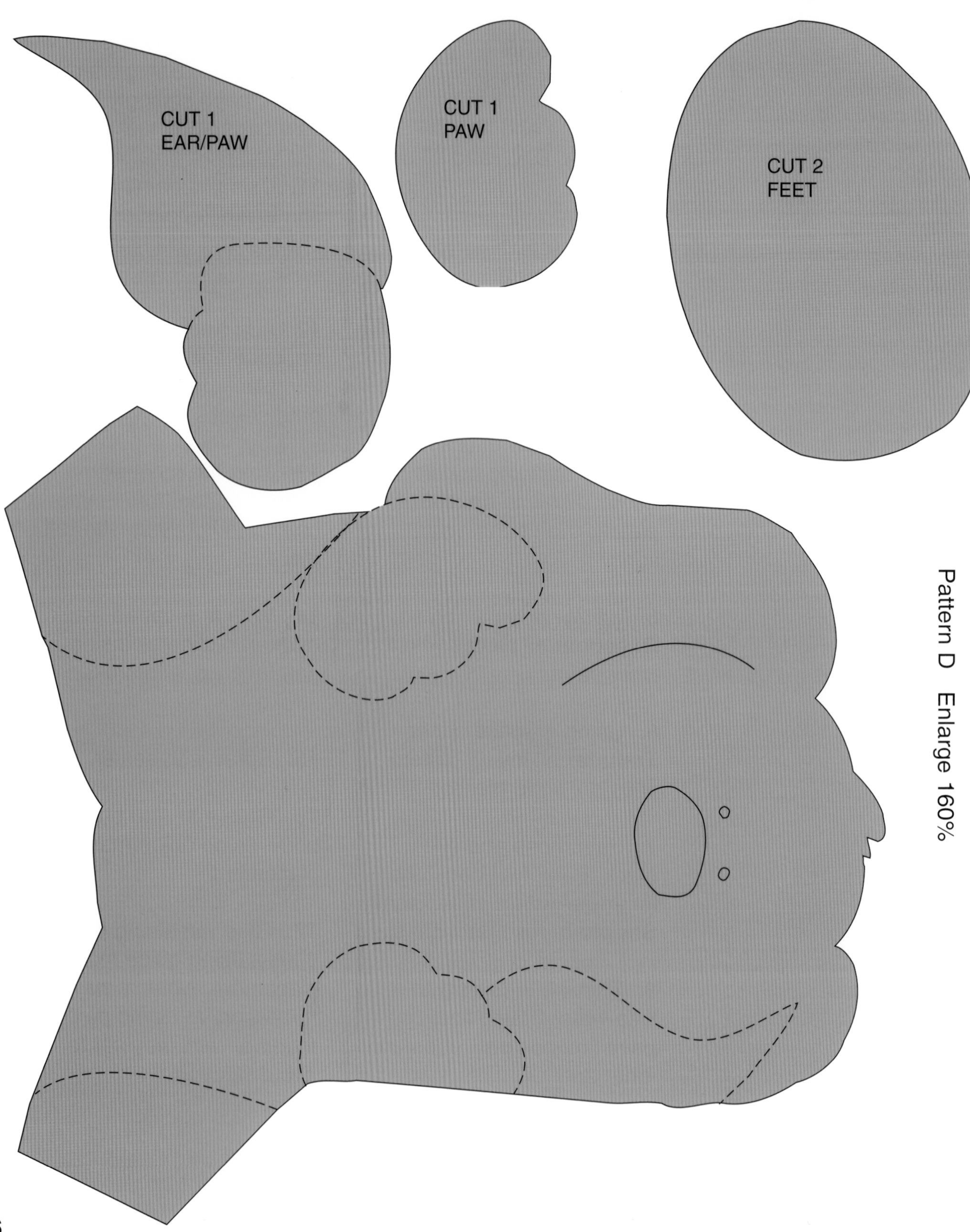
Pattern A Enlarge 160%
Pattern B Enlarge 160%
Pattern C Enlarge 160%
CUT 1
EAR/PAW
CUT 1
PAW
CUT 2
FEET
Pattern D Enlarge 160%

Chair Vanity

Chair Vanity

Photo on page 77.

Materials

- Acrylic gesso
- Acrylic paints: med. blue, med. pink, turquoise, lt. yellow, white
- Dried flowers: as desired
- Fabric: coordinating, 60"-wide (2 yds.)
- Footstool: small, old
- Mirror: cut to fit frame
- Oval frame (to fit between chair rungs)
- Spray sealer: matte
- Stenciling sponge: foam, small
- Terra-cotta flowerpots: 3" (3)
- Thread: coordinating
- Wood chair

Tools & Supplies

- Craft knife
- Glue: industrial-strength; hot glue gun and glue sticks
- Needles: sewing
- Paintbrushes: small, round; ¼"-wide, flat; ½"-wide, flat
- Pencil
- Sandpaper: fine grit
- Sewing machine
- Staple gun with staples

Instructions

Refer to General Instructions on pages 6-13.

1. If necessary, remove rung from back of chair to fit frame. If chair has been previously painted, sand until smooth. Seal chair with acrylic gesso.

2. Referring to photo, paint each section with solid color. Let dry. Paint designs on chair as desired. Model uses checkerboards, stripes, squiggles, dots, x's and v's.

3. To paint stem pattern, use white paint to paint line up and around one leg. **Transfer** Pattern A to stenciling sponge. Cut leaf stencil from sponge with craft knife. Dip leaf into white paint and stamp along stem, changing positions with each application.

Pattern A   Actual size

4. Paint oval frame and all five pots white. Add designs to each to match chair.

5. Seal wood pieces with acrylic gesso.

6. **Secure Mirror** into oval frame. Secure frame in between rungs on chair with industrial-strength glue.

7. Cut piece of fabric to overlap over top of footstool. Staple in place around sides, pulling fabric tight.

8. To cover bottom of footstool, measure height and add 4". Cut this measurement, using full 60"-width of fabric.

9. Sew a **Circular Ruffle**. Sew 2" hem along bottom of fabric. Fold top edge back 1¼" and gather stitch 1" down from top. Pull gathering threads until ruffle fits around stool. With right sides together, sew short ends together to form circular ruffle. Attach ruffle around footstool with hot glue.

10. Cut remaining fabric in three 4" x 16" lengths. Fold fabric down 1". Gather-stitch across folded edge. Secure by knotting. Place selvage edges in bottoms of pots Attach gathered fabric to pots with hot glue.

11. Place dried flowers in flower pots and place them on chair.

Blooming Mirrors

Blooming Mirrors

Photo on page 79.

Materials

- Acrylic gesso
- Acrylic paints: lt. blue, seafoam green, peach, pink, white, med. yellow
- Dowel ¼"-dia. (3')
- Mirrors: oval, 1½" (3)
- Spanish moss
- Styrofoam block: 4"
- Terra-cotta pot: 4"
- Wood: pine: 16" x 4" x ½"

Tools & Supplies

- Drill and ¼" bit
- Glue: industrial-strength; hot glue gun and glue sticks
- Jigsaw
- Graphite paper: gray
- Paintbrushes: ¼"-wide, flat; 1"-wide, flat; spouncer
- Paper towels
- Sandpaper: fine grit
- Tracing paper

Instructions

Refer to General Instructions on pages 6-13.

1. Enlarge Pattern A, B, and C from page 81. Trace patterns onto wood.

2. Cut out wood patterns with jigsaw. Cut one of Pattern A, three of Pattern B, and two of Pattern C.

3. Sand wood until smooth.

4. Seal wood with acrylic gesso. Allow to dry completely, then sand again lightly.

5. **Basecoat** all leaves and dowel with seafoam green paint using spouncer.

6. **Basecoat** one flower lt. blue, one flower peach, and one flower med. yellow.

7. Paint designs (squiggles, curly cues, dots) onto flowers in contrasting colors using ¼" paintbrush.

8. Thin white paint with water. Use 1" paintbrush to **Wash** thinned paint onto flowers and leaves, then wipe off with paper towels.

9. Distress leaves and flowers by sanding edges until raw wood shows through.

10. Using hot glue gun, attach styrofoam block into terra-cotta pot.

11. Drill holes in back center of flowers and back top center of leaves. Holes should be drilled at angle almost parallel with surface of wood.

12. Using industrial-strength glue, attach mirrors to front center of each flower.

13. Cut dowel into sections. Dowels for leaves should be just long enough so that top of leaves rest on lip of terra-cotta pot. Vary length of other sections so that flowers are at different heights.

14. Using industrial-strength glue, attach flowers and leaves to appropriate dowel sections.

15. Position leaves and flowers into Styrofoam and glue into place. Fill terra-cotta pot with Spanish moss.

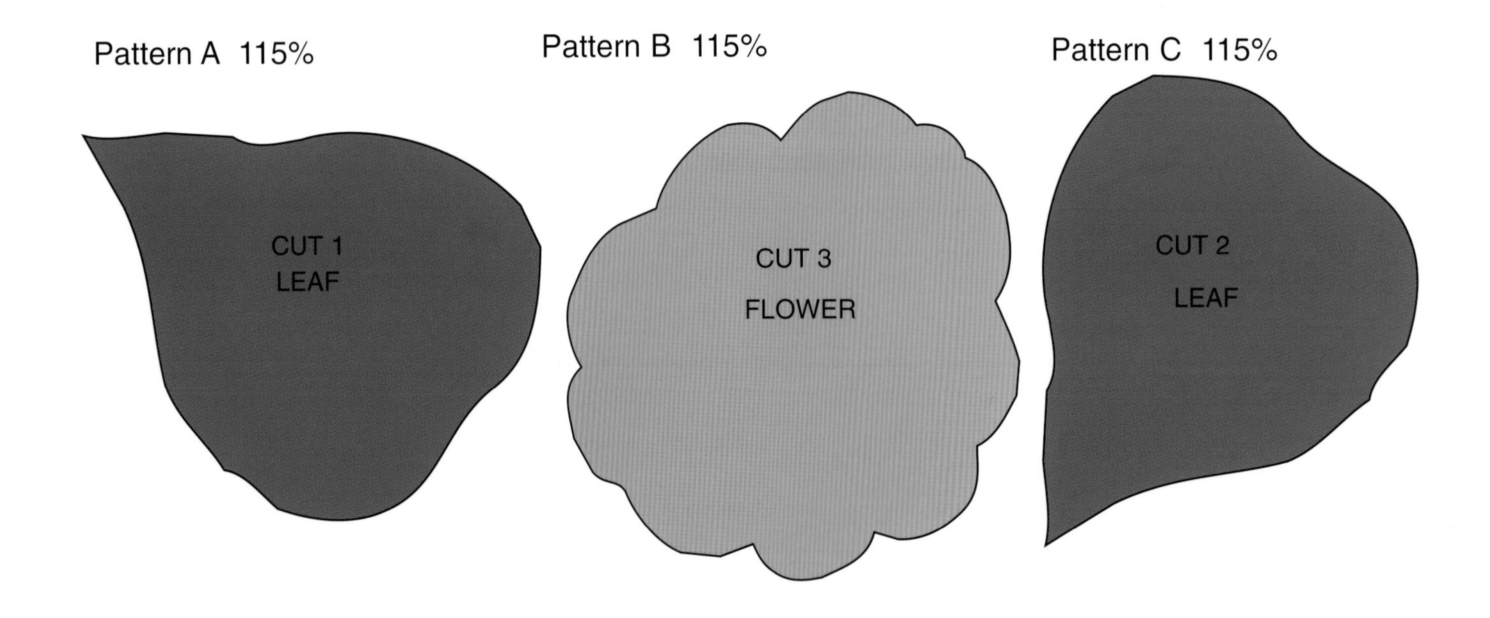

Antique Balusters

Photo on page 82.

Materials

- Acrylic gesso
- Acrylic paint: burnt umber, antique white
- Antiquing liquid: black
- Baluster: 6"-dia. (2)
- Crackle medium
- Mirror: cut to fit frame
- Spray sealer: matte
- Wood frame: pine, basic, square, 3"-thick

Tools & Supplies

- Glue: wood
- Paintbrushes: assorted
- Jigsaw

Instructions

Refer to General Instructions on pages 6-13.

1. Make a Basic Frame or used purchased frame.

2. Seal with acrylic gesso.

3. Cut each baluster in half lengthwise using jigsaw. Place balusters on frame to decide on desired design. See photo on page 82.

4. Cut balusters to fit frame. If unfinished balusters are being used, seal with acrylic gesso.

5. Mix black antiquing liquid in a 1:4 ratio with burnt umber paint. Paint frame and baluster.

6. Following manufacturer's instructions, paint frame and baluster with **Crackle Medium.**

7. Paint frame and baluster with antique white paint.

8. Using wood glue, attach balusters to front of frame in desired design.

9. Spray with sealer

10. Complete by **Securing Mirror** into frame.

Antique Balusters

Window Box Garden

Photo on page 83.

Materials

- Acrylic paints: lt. blue, lt. coral, lt. green, dk. tan, white, lt. yellow
- Battenburg lace doilies: 10" square (3)
- Flowerpots: 6" (3)
- Mirror: 28" x 19"
- Spray sealer: matte
- Wood board: 10' x 1¼" x ½"; 5' x 8" x ½"

Tools & Supplies

- Drill and 1/16"bit
- Glue: wood
- Graphite paper
- Jigsaw
- Paintbrushes: assorted flat; liner
- Paper towels
- Pencil
- Screwdriver
- Wood screws: 1½" (10)

Window Box Garden

Instructions

Refer to General Instructions on pages 6-13.

1. From 10' wood, cut following measurements: two 19½" lengths with mitered ends, two 28½" lengths with mitered ends, two 7⅞" lengths, and one 26" length.

2. From 5' board, cut one piece 7" x 21", one piece 5¼" x 21", and two pieces following Diagram A.

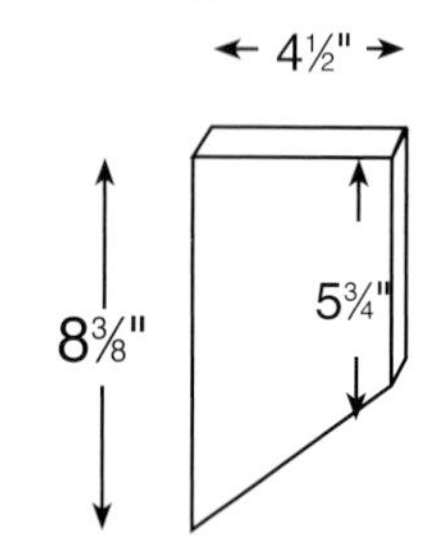

Diagram A

3. Assemble window frame following Diagram B using wood glue.

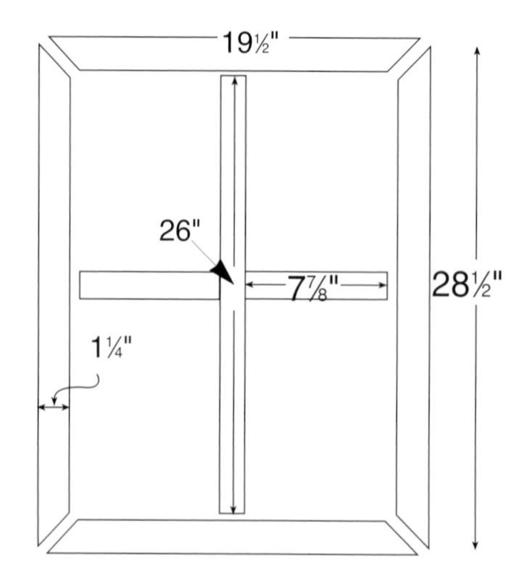

Diagram B

4. Assemble and attach window box to frame as in Diagram C. Screw on bottom, then sides, then front of box. Using drill and 1/16" bit, drill pilot holes for wood screws. Secure wood screws with screwdriver.

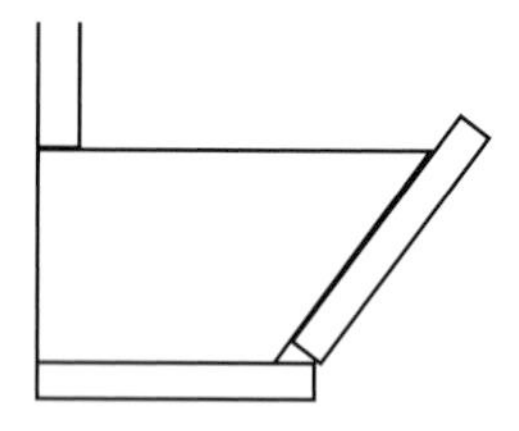

Diagram C

5. Basecoat window box lt. blue. Paint one flowerpot lt. yellow, one lt. green, and one lt.coral.

6. Wash all pieces with thinned white paint. Blot with paper towel for marbled look.

7. Enlarge Pattern A. Using graphite paper, **Transfer** lettering onto frame.

8. Paint over letters with dk. tan paint, using liner brush.

9. Spray with sealer.

10. Glue doilies across top of window frame.

11. Complete by **Securing Mirror** into frame.

Pattern A Enlarge 210%

The tiniest garden is often the loveliest

Barndoor

Photo on page 86.

Materials

- Acrylic gesso
- Acrylic paint: barn red
- Brass hinges: 1" (4)
- Mirror: 8¾"x 9¼"
- Spray sealer: matte
- Wood: pine, 8" x 8" x ½"

Tools & Supplies

- Acrylic gesso
- Dill and 1/16" bit
- Hammer
- Jigsaw
- Nails
- Paintbrush: 1"-wide, flat
- Sandpaper: fine grit
- Screwdriver
- Tack cloth

Instructions

Refer to General Instructions on pages 6-13.

1. From wood, cut six pieces 14½" x 2¼"; cut four pieces 5" x 1¼"; cut two pieces 14½" x 3"; and cut two pieces 13¾" x 3" using jigsaw.

2. Make a Basic Frame, using four 3"-wide pieces.

3. Seal with acrylic gesso.

4. Place three 14½" x 2¼" pieces side by side. Brace together by nailing 5" strip across top and bottom. See Diagram A.

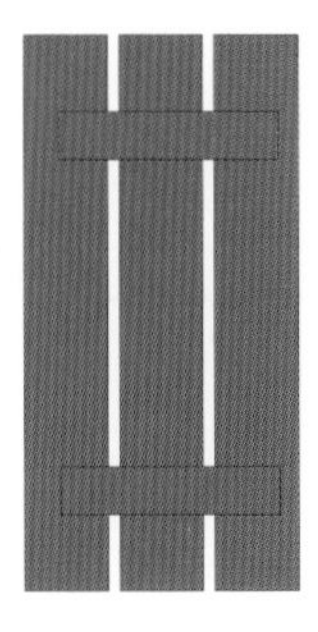

Diagram A

5. Repeat for other side of door.

6. Using drill and screwdriver, attach two hinges to each side of frame to create barn door. See Diagram B.

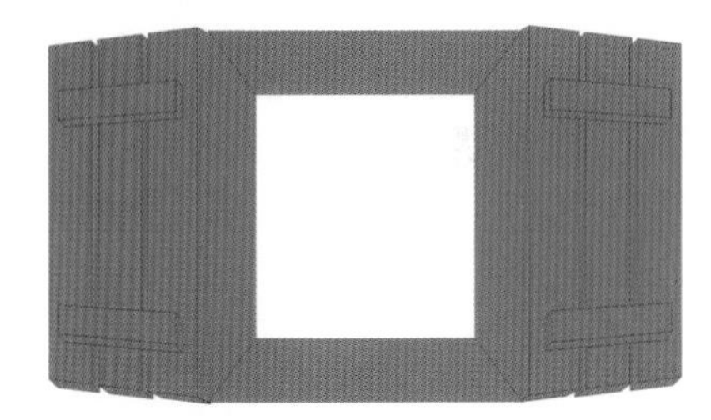

Diagram B

7. Paint all wood barn red.

8. Sand edges of frame and doors lightly to distress. Wipe with tack cloth to remove dust.

9. Spray with sealer.

10. Complete by **Securing Mirror** into frame.

Barndoor

Pocket Mirrors

Pocket Mirrors

Photo on page 87.

Materials

- Acrylic gesso
- Acrylic paints: cream, dk. cream, beige, black, slate blue, gold, ivory, orange, red, silver, dk. tan
- Craft chain (7')
- Eye screw: small (8)
- Jump rings: single (8)
- Lace: scrap
- Mirrors: 2" (4)
- Wood: pine, ¼"-wide, 4" square (4)

Tools & Supplies

- Drill and ⅟₁₆" bit
- Glue: wood
- Paintbrushes
- Pencil
- Pliers
- Scroll saw
- Wire cutters

Refer to General Instructions on pages 6-13.

Lace Purse Instructions

1. Trace desired Pattern A or B on page 89 onto 4" square piece of wood. Cut pattern using scroll saw.

2. Seal with acrylic gesso.

3. **Basecoat** wood with ivory paint.

4. Put lace over purse and stencil through lace with dk. cream paint.

5. Paint frame and clasp of purse with gold paint.

6. Using drill, attach one eye screw ⅟₁₆" in from top edge on each side of purse.

7. Cut craft chain to desired length with wire cutters. Using pliers, open two jump rings. Slip cut chain onto ring. Slip jump ring onto eye screw. Close jump ring with pliers.

8. Glue mirror in center of purse.

Striped Purse Instructions

1. Follow steps 1-2 for Lace Purse.

2. **Basecoat** purse with cream paint.

3. Using slate blue paint, paint thick and thin stripes. See photo on page 87.

4. Paint frame and clasp of purse with silver paint.

5. Follow steps 6-8 for Lace Purse.

Tapestry Purse Instructions

1. Follow steps 1-2 for Lace Purse.

2. **Basecoat** wood with orange paint.

3. Using pencil, draw design as shown in Pattern C.

Pattern C

4. Paint in red areas as shown in photograph with red paint. Paint fine lines with black paint. Do not make fine black lines too straight.

5. Paint frame and clasp of purse with gold paint.

6. Follow steps 6-8 for Lace Purse.

Herringbone Purse Instructions

1. Follow steps 1-2 for Lace Purse.

2. **Basecoat** wood with beige paint.

3. Using pencil, draw lines downward as shown in Pattern D.

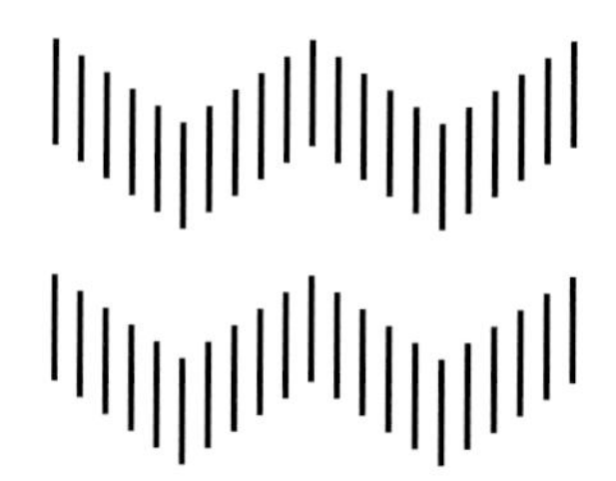

Pattern D

4. Zig-zag short lines made in step 3. Use dk. tan and cream paint to fill in lines.

5. Paint frame and clasp of purse with silver paint.

6. Follow steps 6-8 for Lace Purse.

Pattern A Actual size

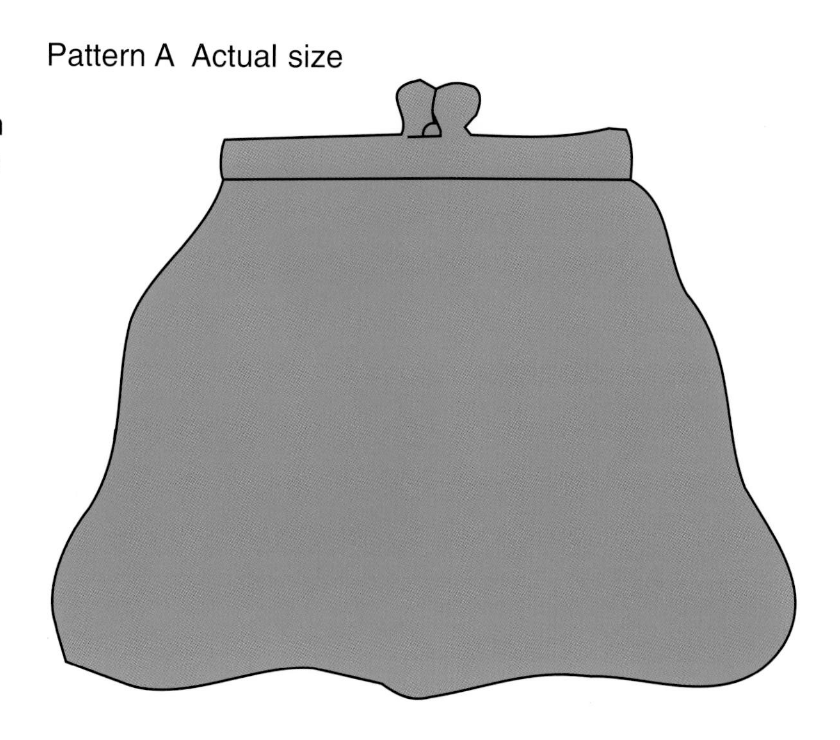

Pattern B Actual size

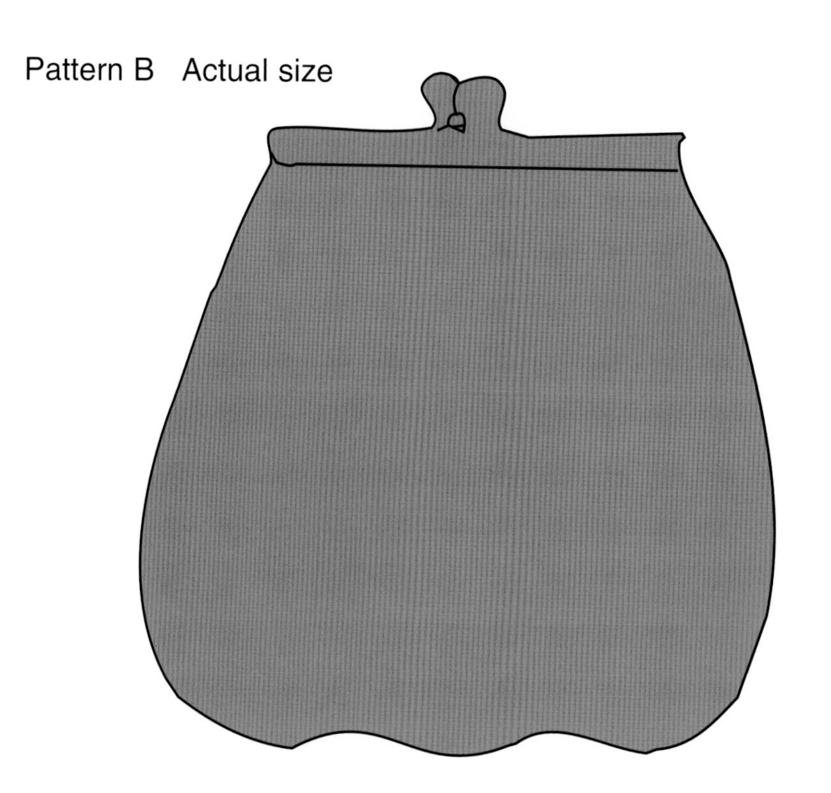

Gallery of Mirrors

Mirrors can light and add interest to homes in more ways than most people imagine. One of a kind mirrors can be made using frames from almost anything imaginable. Trips to antique shops, auctions, flea markets, and garage sales may turn up just the right materials. Look around the house, there may be things there that would be just perfect for creating mirror masterpieces.

Beaded Mirror

A plain oval mirror covered in beads is transformed into a treasured heirloom. Rows of beads were threaded on wire and then attached to a purchased wire wreath. The result of the mirror is unusual yet traditional.

Shaker Mirror

Even the way a mirror is hung can be left up to the imagination. There is no reason a mirror must hang flush against a wall. This mirror is placed on an easel hung on a Shaker peg board as part of a simple grouping.

Harness Mirror

Learn to look for the unusual when choosing mirror frames. With a mirror specially cut to fit, this harness makes a striking accent piece perfect for a rustic decorating scheme.

Daffodil Mirror

This mirror was originally an old faded plastic decorator frame. With a little paint and the addition of a mirror, this flea market find was given a whole new life.

Market Mirror

Mirrors can be as small or as large as desired. This full length mirror is framed with picture frame molding. Specialty framing shops can frame mirrors to fit any space.

Victorian Mirrors

Easy-to-find square mirror tiles can be framed and embellished to give beautiful accents to interior design. Dried or silk flowers and decorative charms can be added to create a Victorian look.

Découpage Roses

A plain flat frame can be dramatically improved with just a few additions. Beveled wood blocks added to the corners and wall paper découpaged on the sides make this mirror extraordinary.

Chapter Four
Shelves

Checkered Heart

Photo on page 97.

Materials

- Acrylic gesso
- Acrylic paints: dk. blue, lt. blue, med. blue, ivory
- Antiquing medium: oil base, brown
- Iron hooks with screw: 4" (3)
- Wood: 22" x 16" x ¾"
- Wood screws: 1½" (2)

Tools & Supplies

- Cloth: lint free
- Dill and ⅛" bit
- Jigsaw
- Paintbrush: ¾"-wide, flat
- Pencil
- Ruler
- Sandpaper: med. grit
- Screwdriver

Checkered Heart

Instructions

Refer to General Instructions on pages 6-13.

1. Enlarge Pattern A. Trace Pattern A onto wood. Cut one shelf; 3¼" x 16". Sand edges smooth.

2. Seal all wood with acrylic gesso.

3. Mark shelf placement on heart. Drill two pilot holes ⅛" in heart. On back of shelf, mark and drill two ⅛" pilot holes to match.

4. Using ivory paint, **Basecoat** entire shelf and heart.

5. Using lines as guide, load blue paints on paintbrush. Paint **Checkerboard Stroke**.

6. Paint outside edge of heart med. blue.

7. Using pilot holes and 1½" wood screws, attach shelf to heart.

8. Using lint free cloth, rub entire project with antique medium.

9. Using screwdriver, attach iron hooks 6" apart.

Pattern A One grid square equals 1"

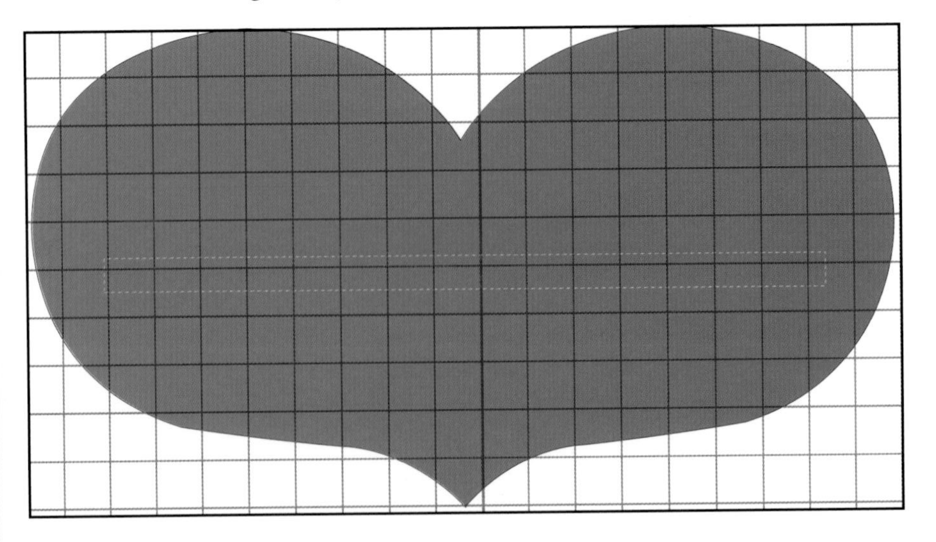

Peach & Pear Shelves

Peach & Pear Shelves

Photo on page 99.

Peach Shelf Materials

- Acrylic paints: dk. brown, dk. green, lt. green, dk. peach lt. peach, yellow
- Plywood: 15" x 10" x ¼"
- Satin finish: clear
- Spray sealer: satin

Tools & Supplies

- Aluminum foil
- Glue: wood
- Jigsaw
- Paintbrushes: ½"-wide, flat; ¼"-wide, flat

Refer to General Instructions on pages 6-13.

Peach Shelf Instructions

1. Enlarge Pattern A and B on page 101. Trace patterns onto wood. Cut peach and shelf using jigsaw.

2. Using aluminum foil for palette, squeeze out quarter size of lt. peach, dk. peach, and yellow paint. Load ½" paintbrush with both peach paints. Following basic shape of peach stroke paint onto peach and shelf.

3. Highlight desired areas of peach with yellow and lt. peach paint. *Note: By blending paint, mistakes are no problem because any color can be added.*

4. Repeat above process for leaf using green paints. Stroke in general shape of leaf adding highlights as desired.

5. Before green paints dry, load ¼" paintbrush with dk. brown paint and stroke onto stem of peach slightly down into leaf.

6. Using wood glue, attach shelf. See photo on page 99.

7. Spray with sealer.

Pear Shelf Materials

- Acrylic paint: dk. brown, med. brown, dk. green, green, lt. green, dk. red-brown, dk. yellow, lt. yellow, med. yellow
- Plywood: 15" x 10" x ¼"
- Spray sealer: satin

Tools & Supplies

- Aluminum foil
- Glue: wood
- Jigsaw
- Paintbrushes: ½"-wide, flat; ¼"-wide, flat

Pear Shelf Instructions

1. Enlarge Pattern B and C on page 101. Trace pattern onto wood. Cut out pear and shelf using jigsaw.

2. Using aluminum foil for palette, squeeze out quarter size of all yellows and browns. Load ½" paintbrush with three yellow paints. Following basic shape of pear, stroke paint onto pear and shelf. Make certain to load yellows often to get all colors on pear.

3. Highlight desired areas by using darker or lighter paint colors.

4. Repeat above process for leaf on pear using green paints. Stroke in general shape of leaf adding highlights as desired.

5. Before green paints dry, load ¼" paintbrush with three brown paints. Stroke stem in slightly down into leaf.

6. Using wood glue, attach shelf. See photo on page 99.

7. Spray with sealer.

Pattern A Enlarge 245%

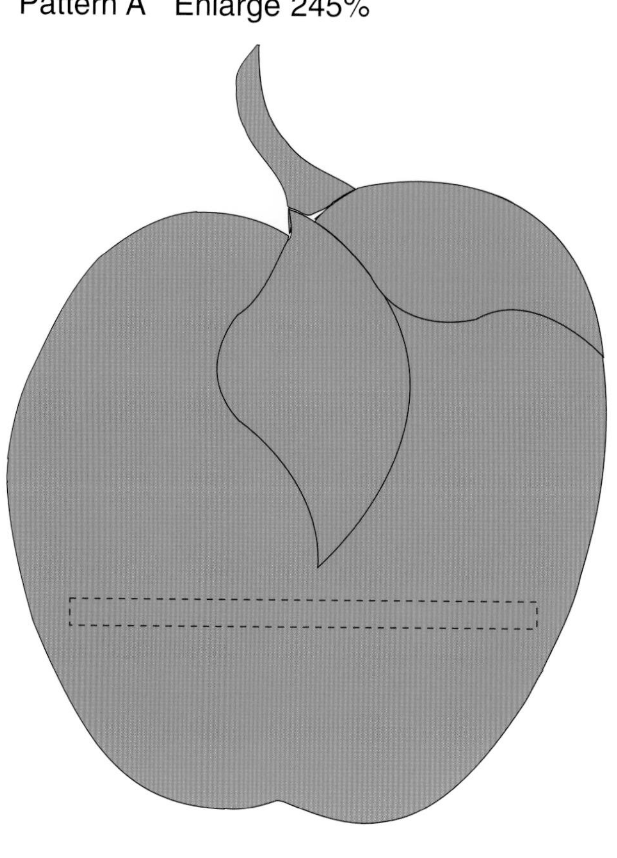

Pattern C Enlarge 245%

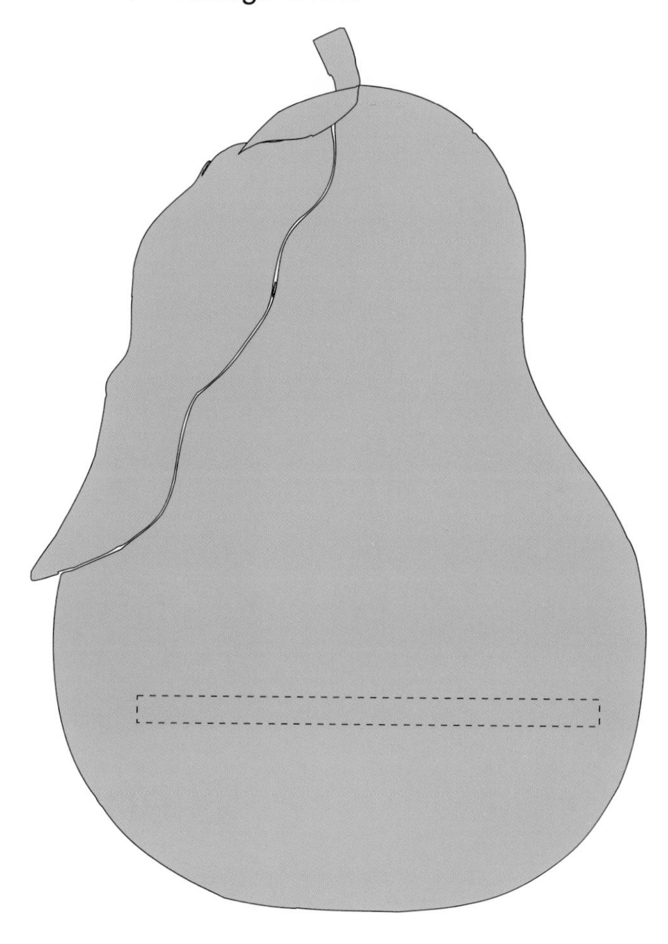

Pattern B Actual size

Blanket Shelf

Materials

- Acrylic gesso
- Acrylic paints: tan, yellow
- Dowel: ⅝" dia. (4')
- Knobs: 2" (2)
- Spray sealer: satin
- Wood: 6" x 8" x 1"; 42½" x 5½" x ¾"

Tools & Supplies

- Drill and 1" bit
- Glue: wood
- Jigsaw
- Paintbrush: ¾"-wide, flat
- Sandpaper: heavy grit; fine grit
- Screwdriver
- Wood putty
- Wood screws: 1½" (14)

Instructions

Refer to General Instructions on pages 6-13.

1. Enlarge Patterns A , B, and C on page 104. Trace pattern onto wood. Cut two ends and one back. Cut one shelf 42½" x 5½" x ¾".

2. Drill holes in both end pieces to accept dowel.

3. Using heavy grit sandpaper, sand top edge of back, front edge of shelf, and front edges of ends.

Make these edges smooth and round.

4. Seal with acrylic gesso.

5. Using wood screws, assemble shelf placing back to end of shelf. See Diagram A. Attach ends with wood screws. Putty holes from screws with wood putty.

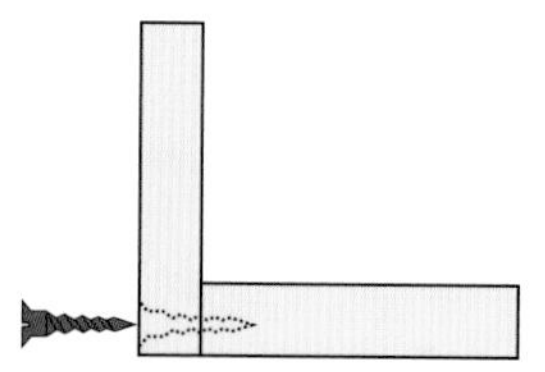

Diagram A

6. Sand shelf smooth with fine grit sandpaper.

7. Using tan paint, **Basecoat** entire shelf, dowel and knobs. Let dry.

8. Using ¾"-wide paintbrush and yellow paint, free-hand yellow stripes over tan base.

9. Insert dowel into two holes drilled in step #2. Using wood glue, glue knobs onto dowels.

10. Spray with sealer.

Pattern A One grid square equals 1"

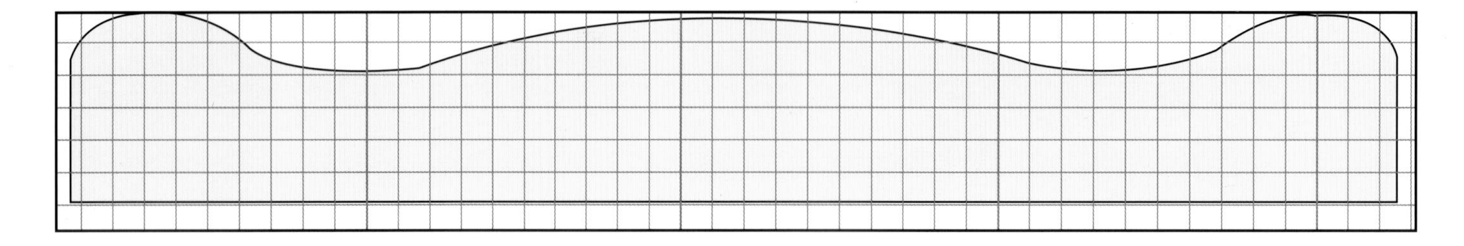

Pattern B One grid square equals 1"

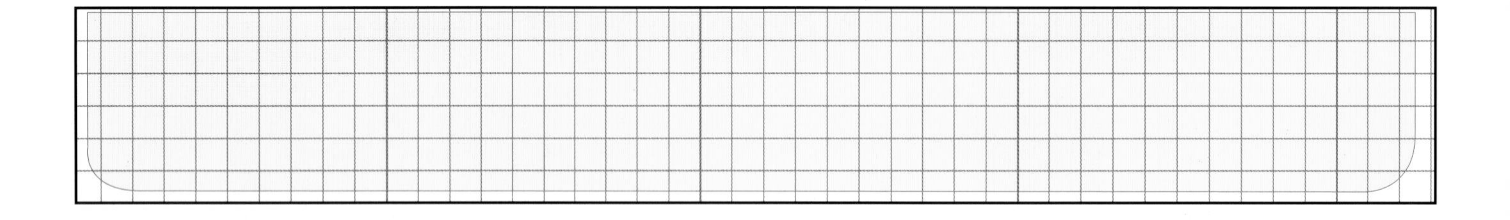

Pattern C Enlarge 145%

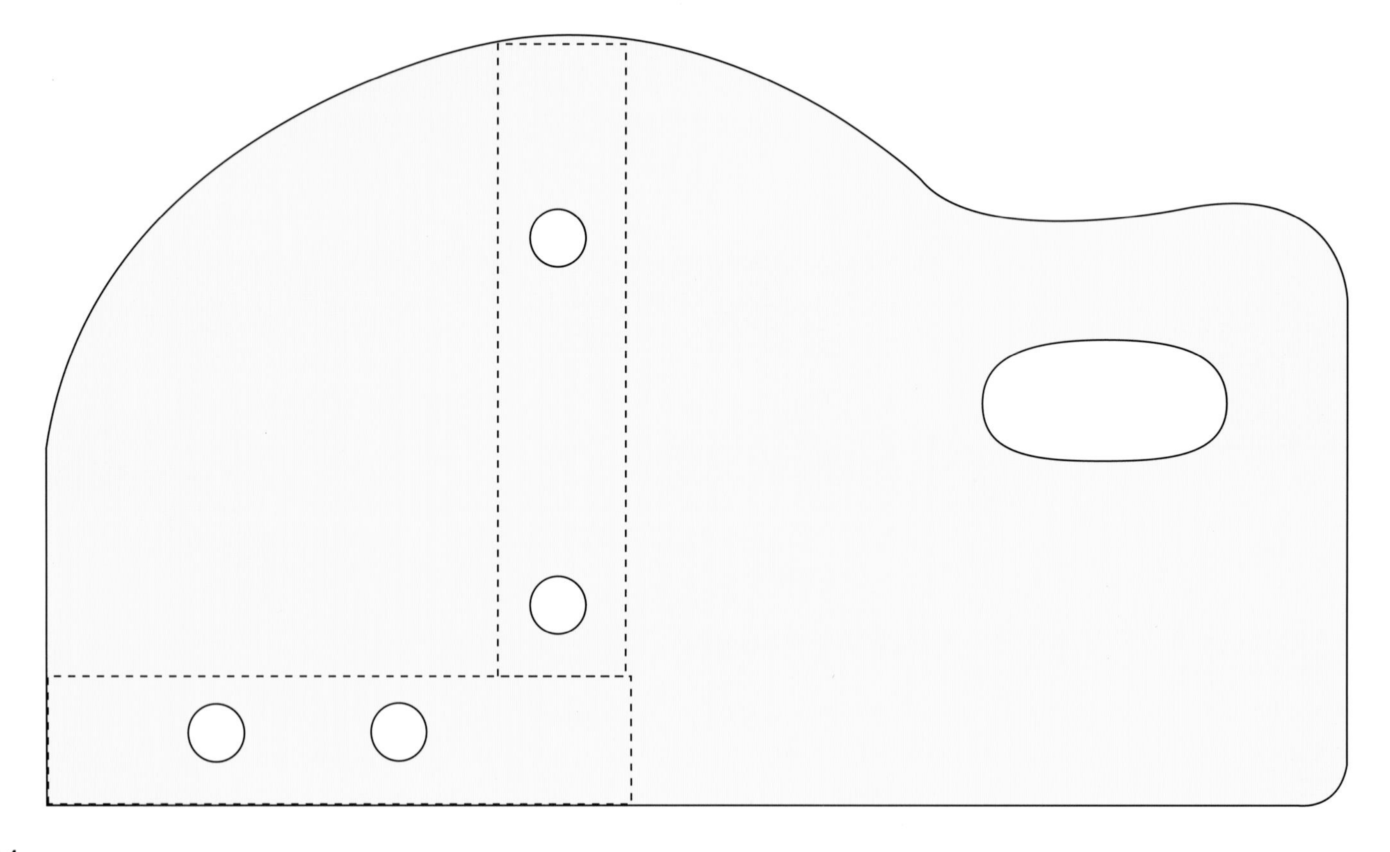

Baluster Shelves

Baluster Shelves

Photo on page 103.

Materials

- Acrylic gesso
- Acrylic paint: as desired
- Balusters: four for each shelf
- Plywood: as desired
- Spray sealer
- Wood screws: 1½" (one for each baluster)

Tools & Supplies

- Drill and 1/16"
- Glue: wood
- Jigsaw
- Measuring tape
- Paintbrushes
- Screwdriver

Instructions

Refer to General Instructions on pages 6-13.

1. From plywood, cut shelves to fit across desired area.

2. Seal shelves and balusters with acrylic gesso.

3. Measure area between desired shelves. Cut balusters to height between shelves.

4. Using drill and screwdriver, place wood screw in top of each baluster that is being attached to shelf. Glue bottom of baluster to shelf.

5. **Basecoat** shelves and balusters as desired.

6. Spray with sealer.

Life, Love, & Light Shelves

Photo on page 108.

Materials

- Acrylic gesso
- Acrylic paints: dk. forest green, ivory, red, dk. yellow
- Crackle medium
- Spray sealer: matte
- Wood: pine, 8" x 18" x ½" (3)

Tools & Supplies

- Drill and ⅛" bit
- Glue: wood
- Paintbrushes: 2"-wide, flat; ½"-wide, flat
- Scroll saw

Instructions

Refer to General Instructions on pages 6-13.

1. Enlarge Pattern A, B, C, and D on page 107. **Transfer** pattern and design onto wood. Cut out each shelf using scroll saw. Drill pilot hole through center of tree design. Cut out design using scroll saw. *Note: To create shelf with heart or star design, place desired enlarged pattern over tree on Pattern A.*

2. Using wood glue, assemble shelves. Seal with acrylic gesso.

3. **Basecoat** front of shelves and back of shelves with desired base color. *Note: Dark colors show up best.*

4. Apply heavy coat of **Crackle Medium**. Make certain to apply crackle medium to sides and ends. Let dry.

5. Slightly thin ivory paint. Using 2" paintbrush, apply ivory paint over crackle medium. Using ½" paint-brush, paint inside of design with ivory paint.

6. Spray with sealer.

Pattern A Enlarge 295%

Pattern B Enlarge 295%

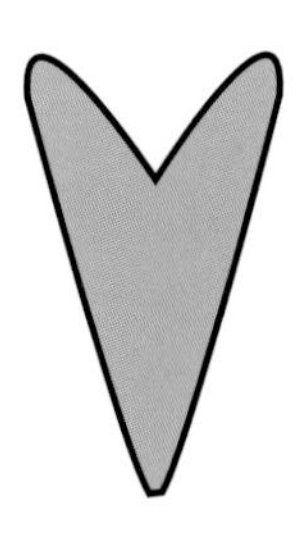

Pattern C Enlarge 295%

Pattern D Enlarge 295%

Knickknack Shelf

Photo on page 109.

Materials

- Acrylic paints: 3 shades of two or three basic colors, cream
- Spice shelf: purchased
- Spray sealer: pearl

Tools & Supplies

- Aluminum foil
- Paintbrush: ⅝"-wide, flat
- Paper towels

Instructions

Refer to General Instructions on pages 6-13.

1. Lay out long piece of aluminum foil, starting with paint from light to dark, squeeze out paints in long row. Keep bottom half of aluminum foil for pallet. *Note: Keep water and paper towels handy. Brush should be cleaned often.*

2. Load paintbrush with colors. **Wash** shelves and inside sides of spice shelf.

3. Paint top, sides, and bottom of shelf using **Basket Weave Stroke**. *Note: The trick to this pattern is to be sloppy. It is not supposed to look perfect.*

4. Spray with sealer.

Life, Love, & Light Shelves

Knickknack Shelf

Moulding Shelf

Materials

- Nails: finishing (2)
- Round moulding: wood, ¼"-wide, same length as shelf
- Wood: 3"-wide, ½"-thick, same length as shelf
- Wood screws: 1½"(8); 2" (4)
- Wood shelf: 3"-wide, ½"-thick, length as desired

Tools & Supplies

- Drill and 1⁄16 bit
- Glue: wood
- Hammer
- Screwdriver

Instructions

1. Measure molding and shelf to fit desired wall space. Cut quarter round moulding same length.

2. Measure, mark, and drill pilot holes along edge of moulding 8 to 12" apart. Repeat pilot holes along back edge of shelf.

3. Using wood screws attach shelf to moulding. See Diagram A.

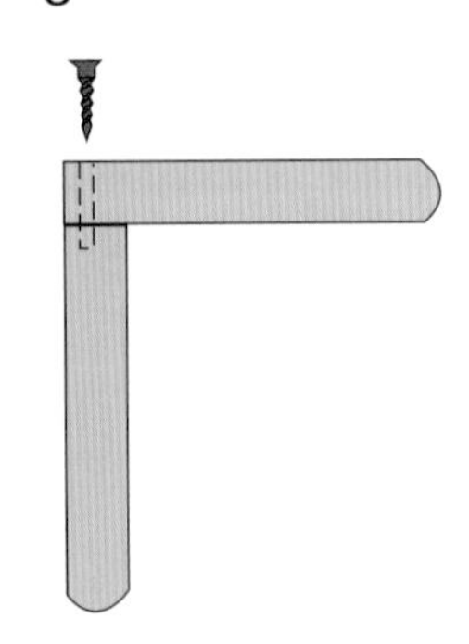

Diagram A

4. Measure and mark placement on wall. Using 2" screw, attach to wall in two places. See Diagram B.

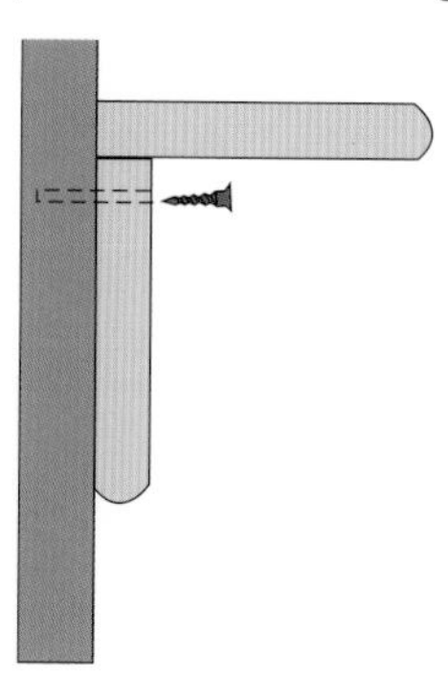

Diagram B

5. Nail quarter round moulding under shelf to hide mounting screws. Place finishing nails into round molding. See Diagram C.

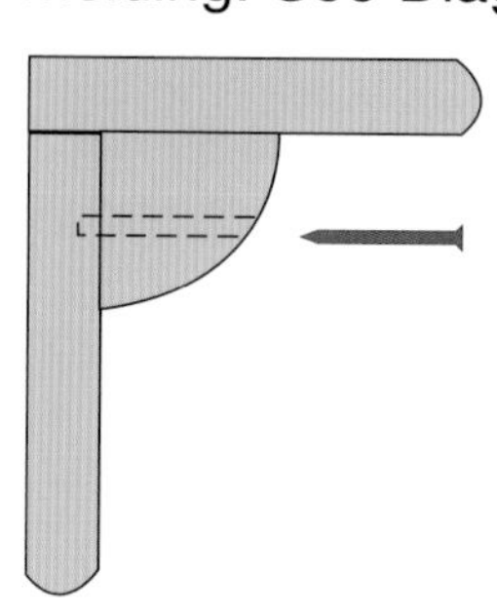

Diagram C

Note: This type of shelf is best for displaying small objects that are lightweight.

Plaid Heart Shelf

Plaid Heart Shelf

Materials

- Acrylic gesso
- Acrylic paints: blue, cream, pink, dk. yellow, lt. yellow,
- Spray sealer: matte
- Wood: 8" x 3" x ⅝" (2)

Tools & Supplies

- Aluminum foil
- Drill and ½" bit
- Glue: wood
- Paintbrushes: ½"-wide, flat; ¼"-wide, flat; liner
- Sandpaper: med. grit
- Scroll saw

Instructions

Refer to General Instructions on pages 6-13.

1. Enlarge Pattern A and B below. **Transfer** patterns onto wood. Cut pattern using scroll saw.

2. Drill pilot hole in center of heart shape. Cut heart shape using scroll saw.

3. Assemble shelf using wood glue. Let dry. Sand edges of shelf. Seal shelf with acrylic gesso.

4. **Basecoa**t shelf with cream paint.

5. Using aluminum foil, place quarter size amount of yellow paints, and pink. Using **Floating** technique and paintbrushes, randomly paint horizontal and vertical lines creating plaid design. Paint one color at a time.

6. **Spatter** shelf with blue paint.

7. Spray with sealer.

Pattern A Enlarge 180%

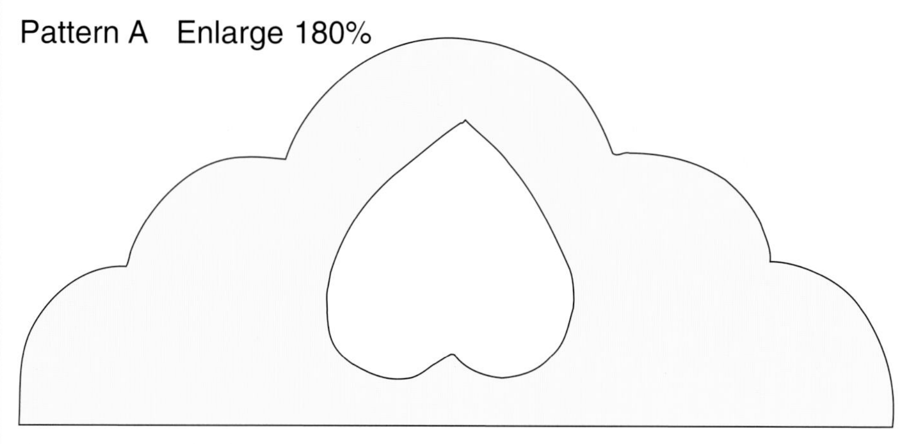

Pattern B Enlarge 180%

Octagon Shelf

Octagon Shelf

Materials
- Acrylic gesso
- Acrylic paints: lt. blue, lt. green, off-white, lt. purple, tan
- Spray sealer: matte
- Nails: 1" (12)
- Wood: 7½" x 3" x ½"

Tools & Supplies
- Drill and 1/16 bit
- Hammer
- Jigsaw
- Paintbrushes: ¾"-wide, flat; fine
- Sandpaper: med. grit
- Wood filler

Instructions

Refer to General Instructions on pages 6-13.

1. Cut eight pieces of wood 6" long cutting ends at 22½° angle. Sand edges.

2. Cut two pieces 12⅞" long. Cut three pieces 3¾" long cutting ends straight. Sand edges.

3. Drill three 1/16" holes across angled ends of 6" pieces. Fitting two angles together, nail into pre-drilled holes. See Diagram A.

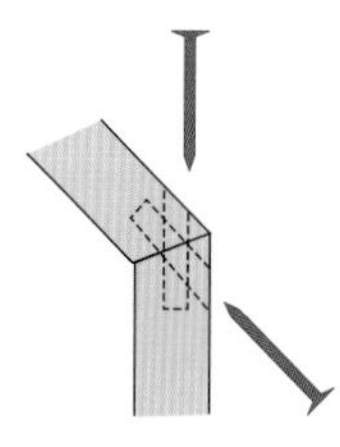

Diagram A

4. Fit all pieces together to form octagon shape. *Note: Due to angles some adjusting of pieces may need to be done to fit all pieces together.*

5. Take two long pieces and two 3¾" pieces. Measure in 4" from each end. Place 3¾" pieces between two longer pieces. Drill holes in long pieces. Nail together to form inner shelves. See Diagram B.

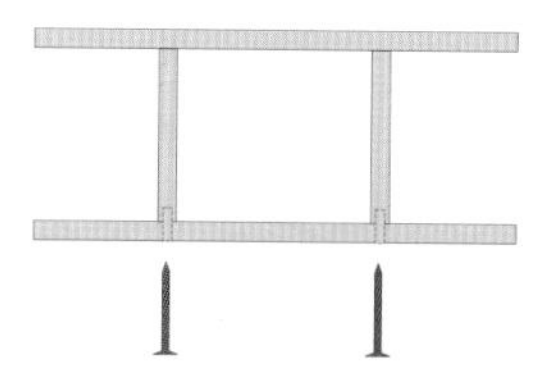

Diagram B

6. Form octagon around pieces formed in step #5. Drill holes where pieces of wood meet and nail together.

7. Center last 3¾" piece of wood at top of long shelf. Drill hole from top of octagon. Nail together to complete octagon. See Diagram C.

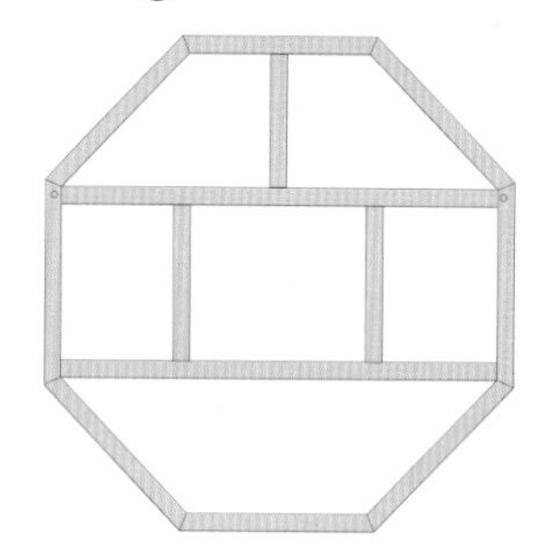

Diagram C

8. Seal with acrylic gesso.

9. **Wash** shelf, painting one square at a time. Use only one color per square. Use **Floating** technique on edges. Freehand word(s) with pencil. Thin paint down to do lettering.

10. Spray with sealer.

Green Picket Shelf

Green Picket Shelf

Materials

- Acrylic gesso
- Acrylic paint: green
- Cedar pickets: pre-cut, 3' x 3½" x ½" (6)
- Nails: 1½" (12)
- Spray sealer: matte

Tools & Supplies

- Hand saw
- Hammer
- Paintbrush: stiff
- Sandpaper: med. grit

Instructions

Refer to General Instructions on pages 6-13.

1. Using hand saw, cut tops off of four pickets at 28" creating shelves.

2. Seal with acrylic gesso.

3. Lightly sand edges of all wood.

4. Measure and mark shelf placements 9" apart on remaining two pickets.

5. Attach one board to bottom placement on picket using nails. Repeat for other three pickets.

6. Line up shelves on second picket. Use nails to attach. See Diagram A.

Diagram A

7. **Basecoat** shelf using green paint

8. Spray with sealer

Virginia Wimmer

In every heart there is a set of shelves designed to shelter only the most precious storage. These shelves are built to hold hopes, dreams, and tender moments. Pack your shelves with gladness everyday, for if left empty for long, these shelves will soon become littered by sadness, fear, and loneliness. Never tie up space on your heart's shelf with troubles. Troubles create clutter any heart is better off without. Toss problems, sorrow, and worry away and in their void pack your heart with sweet memories to relive another day.

Copper Shelves

Copper Shelves

Materials

- Acrylic gesso
- Acrylic paints: dk. brown, orange-brown, umber
- Adhesive: gold leaf
- Leafing: copper, gold
- Matte medium
- Oil paint: burnt sienna
- Paint thinner
- Spray sealer: satin
- Tiles: 12" x 12", dk. brown, red brown, umber
- Tile Grout
- Wood: pine, 12" x 12" x ½" (2)

Tools & Supplies

- Glue: industrial-strength
- Hammer
- Jigsaw
- Paintbrush: ½"-wide, flat; old
- Sponge

Refer to General Instructions on pages 6-13.

Tile Shelf Instructions

1. Enlarge Pattern A, B, and C from page 120. Trace patterns onto wood. Cut design using jigsaw. Assemble shelf using industrial-strength glue.

2. Seal with acrylic gesso.

3. **Basecoat** shelf dk. brown doing a section at a time. With paint still wet, streak umber paint in downward motion. Do not paint shelf top.

4. Using hammer, break tiles in small pieces. Glue tile pieces on like mosaic, placing pieces fairly close together.

5. Following manufacturer's instructions mix tile grout. Add orange-brown paint to grout mixture. Mix well. Smear grout over tiles, making certain to get enough in between tiles.

6. Using wet sponge wash off top of tiles. Rinse sponge out often. An old paint brush can also be used to get grout off of tiles. Let grout dry.

7. Using matte medium, paint over grout and tile to seal. Let dry.

8. Using gold leaf adhesive, cover edge of shelf and grout. Let dry for one hour.

9. Cover shelf edge with copper leafing. Cover grout edge of shelf with gold leafing. Brush excess leafing off lightly.

10. Using paint thinner, thin down oil paint. Paint thin coat over leafing.

11. Spray shelf with satin sealer.

Copper Leaf Shelf Instructions

1. Enlarge Pattern D, E, and F on page 120. Trace patterns onto wood. Cut design using jigsaw. Assemble shelf using industrial-strength glue.

2. Seal with acrylic gesso.

3. **Basecoat** shelf dk. brown doing a section at a time. With paint still wet, streak with umber paint in an outward motion. Cover entire shelf streaking with umber paint on top of shelf.

4. Apply gold leaf adhesive. onto top of shelf, lip of shelf, and bracket. See Diagram A on page 120. Let dry for one hour. Alternating small pieces of gold and copper leafing, over glued areas.

Diagram A

5. Lightly brush off excess leafing.

6. Using paint thinner, thin oil paint. Paint thin coat over leafing.

7. Spray shelf with satin sealer.

Pattern A Enlarge 280%

Pattern C Enlarge 375%

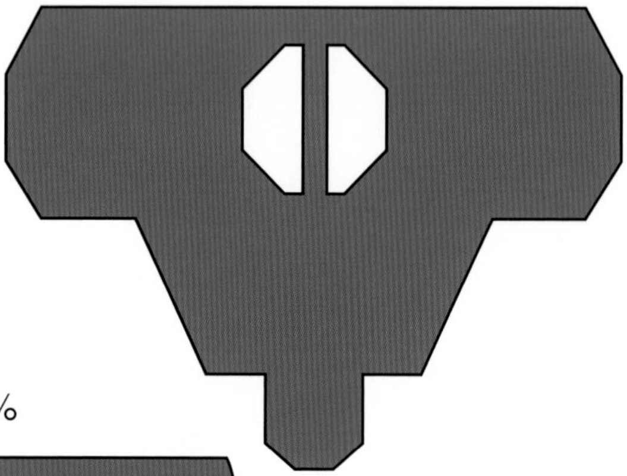

Pattern B Enlarge 290%

Pattern D Enlarge 365%

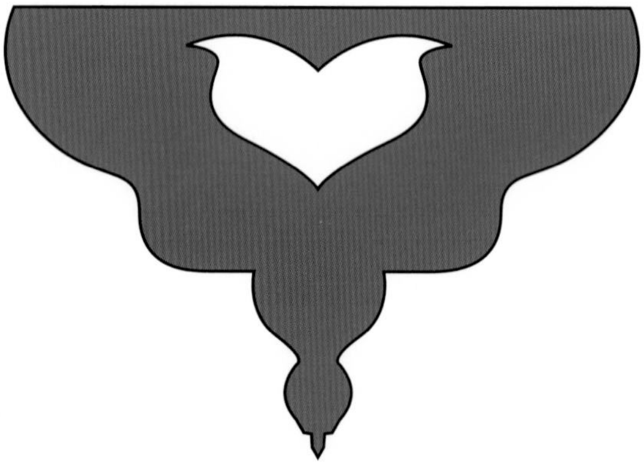

Pattern E Enlarge 325%

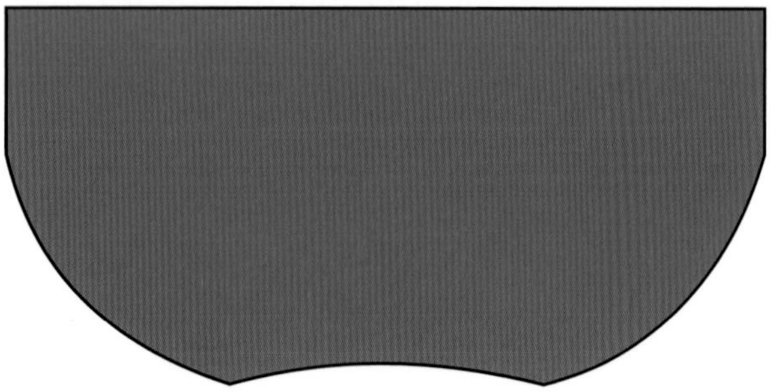

Pattern F Enlarge 315%

Gallery of Shelves

Shelves are a necessity in every home. Whether used to display decorator items, or to house a precious miniature collection, shelves do not have to be expensive or store bought. With a little bit of ingenuity, ideas for shelves can be found anywhere.

Miniature Chair Shelves

Doll-sized chairs hung on a wall make charming shelves for any room. Chairs can be painted in a variety of colors to coordinate with walls or furniture. Holding tiny porcelain tea sets, or small framed pictures these chairs would be perfect in a kitchen or little girl's room.

Drawer Shelf

This arrangement shows that just about anything can be transformed into shelves. This old drawer was made into a decorative shelf to hold candles with the addition of two pieces of wood cut to fit the width of the drawer. Pegs have been added so that candles may hang.

Bench Shelves

Benches can be stacked to form a simple shelf arrangement. Since the benches are attached to each other by gravity only, the arrangement can be changed quickly and easily.

Creative Crates

Shelves can convey as much nostalgia as their contents. Soda bottle crates, available in many antique stores, can be used as instant shelves for miniature displays.

Window Frame

Look carefully before throwing out anything. There may be a potential shelf hiding in that trash. This set of shelves was created using an old window frame and window box.

Planter Box

This chicken wire covered planter box has two shelves cut to fit inside of it. Placed on a worn dresser it adds to the rustic look of this bedroom.

Metric Equivalency Chart

mm-millimetres cm-centimetres
inches to millimetres and centimetres

inches	mm	cm	inches	cm	inches	cm
⅛	3	0.3	9	22.9	30	76.2
¼	6	0.6	10	25.4	31	78.7
½	13	1.3	12	30.5	33	83.8
⅝	16	1.6	13	33.0	34	86.4
¾	19	1.9	14	35.6	35	88.9
⅞	22	2.2	15	38.1	36	91.4
1	25	2.5	16	40.6	37	94.0
1¼	32	3.2	17	43.2	38	96.5
1½	38	3.8	18	45.7	39	99.1
1¾	44	4.4	19	48.3	40	101.6
2	51	5.1	20	50.8	41	104.1
2½	64	6.4	21	53.3	42	106.7
3	76	7.6	22	55.9	43	109.2
3½	89	8.9	23	58.4	44	111.8
4	102	10.2	24	61.0	45	114.3
4½	114	11.4	25	63.5	46	116.8
5	127	12.7	26	66.0	47	119.4
6	152	15.2	27	68.6	48	121.9
7	178	17.8	28	71.1	49	124.5
8	203	20.3	29	73.7	50	127.0

yards to metres

yards	metres	yards	metres	yards	metres	yards	metres	yards	metres
⅛	0.11	2⅛	1.94	4⅛	3.77	6⅛	5.60	8⅛	7.43
¼	0.23	2¼	2.06	4¼	3.89	6¼	5.72	8¼	7.54
⅜	0.34	2⅜	2.17	4⅜	4.00	6⅜	5.83	8⅜	7.66
½	0.46	2½	2.29	4½	4.11	6½	5.94	8½	7.77
⅝	0.57	2⅝	2.40	4⅝	4.23	6⅝	6.06	8⅝	7.89
¾	0.69	2¾	2.51	4¾	4.34	6¾	6.17	8¾	8.00
⅞	0.80	2⅞	2.63	4⅞	4.46	6⅞	6.29	8⅞	8.12
1	0.91	3	2.74	5	4.57	7	6.40	9	8.23
1⅛	1.03	3⅛	2.86	5⅛	4.69	7⅛	6.52	9⅛	8.34
1¼	1.14	3¼	2.97	5¼	4.80	7¼	6.63	9¼	8.46
1⅜	1.26	3⅜	3.09	5⅜	4.91	7⅜	6.74	9⅜	8.57
1½	1.37	3½	3.20	5½	5.03	7½	6.86	9½	8.69
1⅝	1.49	3⅝	3.31	5⅝	5.14	7⅝	6.97	9⅝	8.80
1¾	1.60	3¾	3.43	5¾	5.26	7¾	7.09	9¾	8.92
1⅞	1.71	3⅞	3.54	5⅞	5.37	7⅞	7.20	9⅞	9.03
2	1.83	4	3.66	6	5.49	8	7.32	10	9.14

Index

Antique Balusters 81-82
Antique Medium 8
Around the World 23-24
Baluster Shelves 105-106
Basecoat 8-9
Basket Weave Stroke 9
Barndoor 85-86
Beaded Mirror 90
Bench Shelves 123
Beveled Glass 55-56
Bird's Nest 30-31
Blanket Shelf 102-104
Blooming Mirrors 79-81
Bunny Mirror 74-76
Chair Vanity 77-78
Chapter One:
Frames 14-45
Chapter Two: Techniques
on Mirrors 46-65
Chapter Three: Unique
Mirror Ideas 66-89
Chapter Four: Shelves
. . . . 97-120
Checkerboard Mirror
. . . . 70-72
Checkerboard Stroke 9
Checkered Heart 97-98
Circular Ruffle 12
Copper Shelves 118-120
Crackle Medium 8
Craft Glue 6
Creative Crates 124
Daffodil Mirror 93
Découpage Glue 6
Découpage Roses 96
Drawer Shelf 122
Dry-brush 9
Eye Screw Hangers 6
Family Heirloom 51-52
Flat 8
Flat Frame 12
Float 10
Floral Mirror 60-62
Fly Fishing 68-69
Gallery of Mirrors 90-96
Gallery of Shelves
. 121-126
Garden Gate 66-67
Garden Wall 31-33
General Instructions 6-13
Gilded Window 53-54
Gluing 6
Green Picket Shelf
. 116-117
Hanging Mirrors 6-7
Harness Mirror 92
Home Tweet Home
. . . . 38-41
Hot Glue Gun and
Glue Sticks 6
Industrial-strength Glue 6
Knickknack Shelf
. . . . .107, 109
Life, Love, & Light
Shelves 106-108
Lighthouse 34-36
Liner 8
Making a Basic Frame 7
Marbleize 10
Market Mirror 94
Metal Moose 25-26
Metallic Chain & Ball
. . . . 18-19
Metric Equivalency Chart . .127
Miniature Chair Shelves
. 121
Moonrise 48-50
Moulding Shelf 110-111
No Place Like Home
. . . . 46-47
Oak Vanity 19-21
Octagon Shelf 114-115
Paintbrushes 7-8
Paint Mediums 8
Painting Techniques 8-11
Peach & Pear Shelves
. . . . 99-101
Pearls & Lace 22-23
Plaid Heart Shelf 112-113
Planter Box 126
Play Ball 16-17
Pocket Mirrors 87-89
Roses & Vines 42-43
Routed Edge Frame 12
Running Stitch 13
Saw-tooth Hangers 7
Scroll Saw Silhouette
. . . . 57-59
Sculpting Clay 11-12
Seashore 37-38
Securing a Mirror 12
Shaker Mirror 91
Soldering 12
Spatter 10
Spiked Mirror 72-73
Sponge Stencil 10-11
Spouncer 8
Square Tiles 28-29
Stipple 11
Stitches 12-13
Sundial 63-65
Swirl 11
Tacky Glue 6
Tea Dyeing 13
Teacher's Busy 43-45
Thimbles & Spools 14-15
Transferring from Grid 13
Transferring Patterns 13
Transferring with Paper 13
Tricks & Tips 13
Victorian Mirrors 95
Wash 11
Window Box Garden . . . 83-85
Window Frame 125
Window Screen 26-27
Wood Glue 6

I look in the mirror through the eyes